Competency-Based Critical Care

Simon V. Baudouin (Ed.)

Sepsis

Simon V. Baudouin, MBBS, MD, FRCP, European Diploma in ITU
Royal Victoria Infirmary and the University of Newcastle-upon-Tyne
Newcastle-upon-Tyne
UK

British Library Cataloguing in Publication Data

Sepsis. – (Competency based critical care)
   1. Septicemia
   I. Baudouin, Simon
   616.9′44

   ISBN-13: 9781846289385

Library of Congress Control Number: 2007927933

ISSN: 1864-9998
ISBN-13: 978-1-84628-938-5   e-ISBN-13: 978-1-84628-939-4

9 8 7 6 5 4 3 2 1

Springer Science+Business Media
springer.com

Contents

Contributors vii

Chapter 1   Sepsis: Introduction and Epidemiology 1
            Simon V. Baudouin

Chapter 2   Mechanisms of Innate Immunity in Sepsis 5
            Stuart F.W. Kendrick and David E.J. Jones

Chapter 3   Metabolic and Endocrine Changes in Sepsis and the
            Catabolic State 11
            Steven G. Ball

Chapter 4   Hematological and Coagulation Changes
            in Sepsis 17
            Tina T. Biss and J. Wallace-Jonathan

Chapter 5   The Genetics of Sepsis and Inflammation 26
            Martin F. Clark

Chapter 6   Cardiac, Circulatory, and Microvascular Changes in
            Sepsis and Multiorgan Dysfunction Syndrome 32
            Chris Snowden and Joseph Cosgrove

Chapter 7   Specific Bacterial Infections in the
            Immunocompetent Patient 40
            Hamad A. Hadi and D. Ashley Price

Chapter 8   Infection in the Immunocompromised Patient 47
            Michael H. Snow and Nikhil Premchand

Chapter 9   Severe Infections in the Returning Traveler 56
            Jacob P. Wembri and Matthias L. Schmid

Chapter 10 Antibiotic Prescribing Including Antibiotic
            Resistance 63
            Debbie Wearmouth and Steven J. Pedler

Chapter 11   Infection Control in the Intensive Care Unit 70
             David Tate and Steven J. Pedler

Chapter 12   Randomized Controlled Trials in Sepsis 78
             Helen J. Curtis and Anna Harmar

Chapter 13   Guidelines, Protocols, and the Surviving Sepsis
             Guidelines: A Critical Appraisal 87
             Ian Nesbitt

Chapter 14   Practical Approaches to the Patient with Severe Sepsis:
             Illustrative Case Histories 93
             Victoria Robson

Index 99

Contributors

Simon V. Baudouin, MBBS, MD, FRCP, European Diploma in ITU
Royal Victoria Infirmary and the University of Newcastle-upon-Tyne
Newcastle-upon-Tyne
UK

Steven G. Ball, BSc, MBBS, PhD, FRCP
Senior Lecturer
School of Clinical Medical Sciences
University of Newcastle
Newcastle-upon-Tyne
UK

Tina T. Biss, MBBS, MRCP, DIP, FRC Path
Department of Haematology
Freeman Hospital
Newcastle-upon-Tyne
UK

Martin F. Clark, MB ChB, BMSC, FRCA, DICM
Department of Anaesthesia
Queen Margaret Hospital
Dunfermline
Fife, Scottland

Joseph Cosgrove, FRCA
Department of Anaesthesia and Critical Care Medicine
Freeman Hospital
Newcastle-upon-Tyne
UK

Helen J. Curtis, MBBS, FRCP
Department of Anaesthesia
Royal Victoria Infirmary
Newcastle-upon-Tyre
UK

Hamad A. Hadi, MBBS, MRCP
Department of Infectious Diseases and Tropical Medicine
Newcastle General Hospital
Newcastle-upon-Tyne
UK

Anna Harmar, MBBS
Department of Anaesthesia
Royal Victoria Infirmary
Newcastle-upon-Tyre
UK

David E.J. Jones, BA, BM, BCh, MRCP, FRCP, CCST (Gastroenterology)
Applied Immunobiology and Transplantation Research Group
School of Clinical Medical Sciences
University of Newcastle
Newcastle-upon-Tyne
UK

Stuart F.W. Kendrick, MA, BM, BCh, MRCP
Applied Immunobiology and Transplantation Research Group
School of Clinical Medical Sciences
University of Newcastle
Newcastle-upon-Tyne
UK

Ian Nesbitt, MBBS(Hons), FRCA, DICM(UK)
Integrated Critical Care Unit
Freeman Hospital
Newcastle-upon-Tyre
UK

Steven J. Pedler, MB ChB, FRC Path
Department of Microbiology
Royal Victoria Infirmary
Newcastle-upon-Tyne
UK

Nikhil Premchand, BSc, MBBS, MRCP,
 DTM&H
Department of Infectious Diseases and
 Tropical Medicine
Newcastle General Hospital
Newcastle-upon-Tyne
UK

D. Ashley Price, MBBS, DTMH, MRCP
Consultant Infectious Diseases
Department of Infectious Diseases and
 Tropical Medicine
Newcastle General Hospital
Newcastle-upon-Tyne
UK

Victoria Robson, BSc, MBBS, MRCP(UK), FRCA,
 EDIC
Consultant (Anaesthesia and Critical Care)
Department of Anaesthesia
Royal Victoria Infirmary
Newcastle-upon-Tyne
UK

Matthias L. Schmid, MD, FRCP, DTM&H
Consultant Physician and Clinical Senior
 Lecturer
Department of Infection and Tropical
 Medicine
Newcastle General Hospital
Newcastle-upon-Tyne
UK

Michael H. Snow, MBBS, FRCP
Consultant Physician
Department of Infection and Tropical Medicine
Newcastle General Hospital
Newcastle-upon-Tyne
UK

Chris Snowden, B. Med. Sci (Hons), MD, FRCA
Department of Anaesthesia and Critical Care
 Medicine
Freeman Hospital
Newcastle-upon-Tyne
UK

David Tate, MB ChB
Department of Microbiology
Royal Victoria Infirmary
Newcastle-upon-Tyre
UK

J. Wallace-Jonathan, MBBS, FRCP(UK),
 FRC Path
Department of Haematology
Freeman Hospital
Newcastle-upon-Tyne
UK

Debbie Wearmouth, MBBS
Department of Microbiology
Freeman Hosptial
Newcastle-upon-Tyre
UK

Jacob P. Wembri, MRCP, MSc, DTM&H
Department of Infection and Tropical Medicine
Newcastle General Hospital
Newcastle-upon-Tyne
UK

1
Sepsis: Introduction and Epidemiology

Simon V. Baudouin

Sepsis is a major health problem in both the industrialized and nonindustrialized world. In industrialized nations, sepsis is one of the most common illnesses in hospitalized patients and is associated with a substantial mortality (Table 1.1). In the nonindustrialized world, sepsis remains one of the main causes of diminished life expectancy.

Sepsis is a condition that is often easy to recognise clinically but much harder to define. In broad terms, it is the inflammatory response to host microbial invasion. This inflammatory response, which has evolved to combat and limit the spread of infection, produces complex immunological, coagulation, and circulatory changes that may progress to a state of organ dysfunction and failure known as septic shock. This issue of competency-based critical care will review the many different aspects of sepsis and its management.

The Epidemiology of Sepsis

The fact that infection can cause circulatory collapse and organ failure has been recognized for several centuries. However, real scientific interest in the condition only developed in the 1960s, when it became apparent that infection, and, in particular, gram-negative infections, were increasingly common among the hospital population. Although early definitions of sepsis and septic shock varied, it was clear that the syndrome was associated with a high mortality, often in excess of 50% in reports.

A lack of consensus regarding the definition of sepsis hampered research in the field and, in 2001, a consensus statement and definitions were published jointly by the North American and European Critical Care Societies [1] (Table 1.1). These had the laudable aim of improving and standardizing recruitment into large clinical trials. In addition, standardization has improved our understanding of the epidemiology of sepsis. However, the definitions have not been without problems. A major criticism has been that, in an attempt to recruit large numbers of patients into clinical trials, several different conditions have been "lumped" together. For example, is it reasonable to consider the clinical phenotype of a previously fit 17-year-old with meningococcal sepsis to be the same as an 81-year-old with chronic ill health and faecal peritonitis?

These issues may be resolved by the development of new, more complex descriptors of sepsis. A recent International Sepsis forum consensus conference developed definitions for the six most common infections causing sepsis, including pneumonia, blood-stream infections, intra-abdominal infections, urological infection, and surgical wound infection [2]. An alternate approach to sepsis definition has been modeled on the tumor size-lymph nodes-metastases (TNM) classification of tumor stage. The PIRO scheme assesses predisposition, infection, response, and organ dysfunction in an attempt to stage sepsis more accurately.

The existence of several competing definitions of sepsis highlights the real difficulty in describing and comparing this condition. Despite some drawbacks, the 2001 Consensus definitions remain an "industry standard" and will only be displaced if newer definitions prove to have greater use.

Incidence and Occurrence of Sepsis

Despite the difficulties with diagnostic criteria, there is good evidence that the incidence and hospital occurrence of sepsis is increasing. The largest study to address these issues was conducted by the North American Centers for Disease Control in 1990. The study reported an increase in sepsis incidence from 73.6 in 100,000 patients in 1979 to 175.9 in 100,000 patients in 1989. The occurrence of sepsis also seems to be rising in the United Kingdom. The Dr Foster Organisation recently reported that admissions for septicemia had increased by 53% between 1996/1997 and 2001/2002 in UK hospitals (Table 1.2). A number of other studies support the finding that sepsis occurrence in hospitalized patients is increasing.

The causes of this rise are not well defined, but an increasingly elderly population, with chronic

TABLE 1.1. Definitions of diseases

Systemic inflammatory response syndrome (SIRS)	Two or more of the following: • Body temperature >38.5°C or<35.0°C • Heart rate >90 beats per minute • Respiratory rate >20 breaths per minute, or arterial CO_2 tension <32 mmHg, or need for mechanical ventilation • White blood cell count >12,000/mm³ or <400/mm³ or immature forms >10%
Sepsis	Systemic inflammatory response syndrome and documented infection (culture or gram stain of blood, sputum, urine, or normally sterile body fluid positive for pathogenic microorganism; or focus of infection identified by visual inspection—e.g., ruptured bowel with free air or bowel contents found in abdomen at surgery, wound with purulent discharge)
Severe sepsis	Sepsis and at least one sign of organ hypoperfusion or organ dysfunction: • Areas of mottled skin • Capillary refilling time ≥3 s • Urinary output <0.5 mL/kg for at least 1 h or renal replacement therapy • Lactate >2 mmol/L • Abrupt change in mental status or abnormal electroencephalogram • Platelet counts <100,000/mL or disseminated intravascular coagulation • Acute lung injury—acute respiratory distress syndrome • Cardiac dysfunction (echocardiography)
Septic shock	Severe sepsis and one of: • Systemic mean blood pressure <60 mmHg (<80 mmHg if previous hypertension) after 20–30 mL/kg starch or 40–60 mL/kg saline, or pulmonary capillary wedge pressure between 12 and 20 mmHg • Need for dopamine >5 µg/kg/min to maintain mean blood pressure above 60 mmHg (80 mmHg if previous hypertension)
Refractory septic shock	Need for dopamine >5 µg/kg/min or norepinephrine or epinephrine >0.25 µg/kg/min to maintain mean blood pressure above 60 mmHg (80 mmHg if previous hypertension)

TABLE 1.2. Top 15 International Classification of Disease (ICD)-10 mortality groups by numbers of deaths, England 2001–2002

No.	ICD-10 mortality group	Percentage of all in-hospital deaths (%)	In-hospital mortality rate (%)
1	Cerebrovascular diseases	10	30.2
2	Pneumonia	9.9	29.7
3	Other heart diseases	8.3	11.2
4	Ischemic heart diseases (including myocardial infarction)	6.2	6.8
5	Myocardial infarction	4.7	16.5
6	Injuries and poisonings	5.4	2.1
7	Malignant neoplasms (other than trachea, etc.; colon, etc.; and breast)	4.6	12.2
8	Malignant neoplasm of trachea, bronchus, and lung	3.8	24.4
9	Chronic lower respiratory diseases	3.5	5.1
10	Remainder of respiratory diseases	2.7	2.8
11	Remainder of circulatory diseases	2.4	4.5
12	Other acute lower respiratory infections	1.9	4.9
13	*Septicemia*	*1.8*	*39.6*
14	Malignant neoplasm of colon, rectum and anus	1.8	9.9
15	Diseases of the liver	1.2	17.2

Source: Adapted from Jarman et al., 2004 [4].

ill health who are undergoing an increasing number of invasive procedures, is likely to be contributing. These changes may also reflect an increasingly proactive approach to the elderly patient with severe illness.

The rising incidence of sepsis has resulted in a significant proportion of hospitalized patients with the condition. Estimates vary widely, but a recent UK survey, based on Intensive Care National Audit and Research Centre (ICNARC) data, found that 27% of adult patients admitted to the intensive therapy unit (ITU) fulfilled severe sepsis criteria in the first 24 hours.

Outcome from Sepsis

All case series and clinical trials of sepsis show that the condition has a very high mortality. Most studies report a hospital mortality rate between 40 and 80% and a recent meta-analysis of 131 studies reported an overall hospital mortality of 50%. Data from the UK ICNARC case-mix program confirms the high mortality [3]. Patients admitted to ITU with severe sepsis had a 35% ITU mortality and a 47% hospital mortality, using information from 1995 to 2000 gathered from 91 UK units. These overall figures mask large variations in outcome. Initial severity of illness is one major determinant of survival. In one multicenter study on sepsis, a stepwise progression in mortality was noted. Patients with systemic inflammatory response syndrome (SIRS) had a 7% mortality, those with sepsis had a 16% mortality, and septic shock was associated with a 46% mortality.

There is some evidence that outcome may be improving in severe sepsis (Figure 1.1). Data from studies between 1960 and 1970 report mortality in excess of 60%. Clinical trials that are more recent have found a lower mortality in the control patient population. For example, 28-day mortality in the control group of the Recombinant Human Activated Protein C [Xigris] Worldwide Evaluation in Severe Sepsis (PROWESS) study was 31%; 34% in the Optimised Phase III Tifacogin in Multicenter International Sepsis trial (OPTIMIST) and 39% in the KyberSept study.

Despite possible improvements in short-term outcome, longer-term survival after severe sepsis

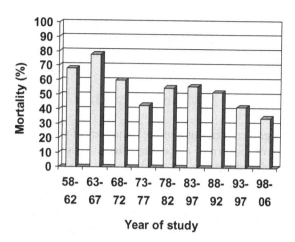

FIGURE 1.1. Trends in mortality from sepsis over time. (Adapted from Angus and Wax, 2001 [5].)

remains poor. This is demonstrated by the follow-up data from the PROWESS study, which reported hospital 3-, 6-, 12-, and 24-month survival, in control patients of 65%, 62%, 60%, 57%, and 49%, respectively.

Predictors of Outcome

A number of patient-related factors have a significant impact on outcome after sepsis. In the short term, increased severity of illness, as measured by Acute Physiology and Chronic Health Evaluation (APACHE) III and the number of failed organs, is associated with poorer outcome. Additional factors associated with increasing mortality are age, the presence of chronic ill health, the site of infection, and the type of organism causing the infection. The importance of individual risk factors may alter with time. Short-term survival is associated with initial severity of illness, but seems less important in predicting longer-term outcome. Increasing age and chronic ill health remain important predictors of poor long-term outcome.

A number of models of outcome have been developed in an attempt to predict outcome in sepsis. These include APACHE III, Simplified Acute Physiology Score (SAPS) II and other, more specific, models. Not surprisingly, none of these functions well when used on groups that were not part of the original data set. The changes over time of risk factors also reduce the likely usefulness of any given model of outcome.

Changes in Microbiology Over Time

Significant changes in the frequency of pathogens have occurred since the 1960s. Original case series of sepsis were dominated by gram-negative infections. Since the mid-1970s, gram-positive infections have become much more important, with methicillin-resistant staphylococcus aureus (MRSA) sepsis now becoming very frequent in hospitalized patients. Fungal sepsis is also becoming more frequent, probably as a consequence of the increasing use of broad-spectrum antibiotics. Changes in pathogen frequency are likely to be the result of the increasing use of intravascular devices; changes in antibiotic presenting practice; increasing ITU admissions of elderly, chronically ill patients; and changes in surgical practice.

The Economics of Sepsis

From an economic perspective, patients with sepsis admitted to critical care units are expensive. This is clear from ICNARC data, which reported that patients with sepsis occupied 45% of all ITU bed days. The major costs of ITU stay are staffing and disposables (including drugs) and both of these costs increase with duration of stay. Absolute costs are difficult to estimate in any healthcare system and comparison of costs between different systems is very difficult. Estimates of median costs per patient from North America vary from $10,000 to more than $20,000. These represent more than six times the average cost of an ITU admission. Nonsurvivors also tend to incur greater costs than survivors because of the greater intensity and duration of treatment.

Conclusion

Sepsis is increasing in frequency in hospitalized patients. The rapid evolution of sepsis to the stage of septic shock provides an immense challenge to the entire acute hospital team. Recognition and treatment of sepsis are core skills for any Intensivist because it is probably the commonest condition that the he or she will manage. Despite some evidence of recent improvements, outcome remains poor and sepsis remains the single most lethal condition in current hospital practice.

References

1. Levy MM, Fink MP, Marshall JC, Abraham E, Angus D, Cook D, Cohen J, Opal SM, Vincent JL, Ramsay G. 2001 SCCM/ESICM/ACCP/ATS/SIS International Sepsis Definitions conference. *Intensive Care Med* 2003;29:530–538.
2. Calandra T, Cohen J. The international sepsis forum consensus conference on definitions of infection in the intensive care unit. *Critical Care Medicine* 2005;33:1538–1548.
3. Gupta D, Keogh B, Chung KF, Ayres JG, Harrison DA, Goldfrad C, Brady AR, Rowan K. Characteristics and outcome for admissions to adult, general critical care units with acute severe asthma: a secondary analysis of the ICNARC Case Mix Programme Database. *Crit Care* 2004;8:R112–R121.
4. Jarman B, Aylin P, Bottle A. Trends in admissions and deaths in English NHS hospitals. *BMJ* 2004;328:855.
5. Angus DC, Wax RS. Epidemiology of sepsis: an update. *Critical Care Medicine* 2001;29:S109–S116.

2
Mechanisms of Innate Immunity in Sepsis

Stuart F.W. Kendrick and David E.J. Jones

The pathogenesis of the sepsis syndrome is critically dependent on activation of the innate immune response. Innate immunity plays a direct role in the development of sepsis and is also crucial for the activation and modulation of later antigen-specific adaptive immune responses. Nearly all of the clinical manifestations of sepsis and the systemic inflammatory response syndrome (SIRS) can be attributed to components of the innate immune response. However, this review focuses on the new and expanding field of innate immune activation by pathogen-responsive receptors, most importantly, the toll-like receptors (TLRs).

The Innate Immune Response in Sepsis

The human innate immune system is the first line of defence against invading pathogens. Its remarkable ability to respond rapidly to a wide range of microorganisms is essential for survival. It combats and contains infection at the point of entry, signals danger to other systems, and allows time for the T and B cells of the more finely tuned, antigen-specific, adaptive immune system to become effective. In evolutionary terms, a vigorous innate immune response will have conferred a survival advantage to our hominid ancestors when the loss of the thick fur characteristic of other primates left the skin vulnerable to frequent injuries and contamination. Although the molecular and cellular components of the innate immune system differ little between mammalian species, the human response to microbial components is one of the most sensitive. The disadvantage of such a vigorous response is an increased susceptibility to exaggerated systemic inflammation and shock when the innate immune system is activated systemically rather than locally.

The innate response is triggered by activation of cells equipped to respond to pathogens or specific pathogen components—cells of the macrophage/monocyte lineage, natural killer cells, dendritic cells, and endothelial cells. The activated cells secrete inflammatory mediators including cytokines (most importantly, tumor necrosis factor [TNF]-α, interleukin [IL]-1, and IL-6), chemokines (such as IL-8), prostaglandins, and histamine. These mediators act on vascular endothelial cells to cause nitric oxide-mediated vasodilatation, increased vascular permeability, and neutrophil recruitment into tissues. The coagulation cascade is activated locally with up-regulation of endothelial tissue factor, and decrease in thrombomodulin and its antithrombotic product, activated protein C.

In systemic sepsis, the local responses become widespread. Systemic vasodilatation causes hypotension, shunting, and reduced tissue oxygen delivery. Endothelial activation and apoptosis result in loss of vascular integrity, proteinaceous exudate, and edema. Disseminated intravascular coagulation produces small-vessel microthrombosis, depletion of clotting factors, and coagulopathy. Reactive oxygen species are generated from activated neutrophils, tissue effects of nitric oxide, and cytokine-induced alterations in cellular metabolism. The cumulative effect of these changes is increasing severity of sepsis, with multiple organ dysfunction and worsening mortality.

Triggering of Innate Immune Responses

Until recently, our understanding of sepsis lacked a description of the mechanism by which cells of the innate immune system recognize and respond to microbial threats. By definition, these cells lack the elegant (but relatively slow and energetically costly) antigen-specific receptor systems that characterize the adaptive immune response. In recent years, our understanding of immune reactivity has changed from the pure discrimination of "self" from "nonself" epitopes to an appreciation of the importance of specific "danger" signals in initiating, directing, and modulating the immune response. Danger signals can come from internal sources that indicate tissue damage or invasion, such as products of cell lysis, coagulation, or complement cascades, or from exogenous material, such as microbial surface molecules or genetic material. It is to these danger signals that the innate immune system responds. The fact that both microbial and internal danger signals can trigger the response explains the similarity of the sepsis syndrome to SIRS with a noninfective precipitant, such as trauma, burns, or pancreatitis.

Recognition of microbial danger signals requires a receptor system that responds to evolutionarily conserved structural components of microorganisms, so that an organism cannot use genetic variability to escape detection. The components that allow microorganisms to trigger the immune response are termed pathogen-associated molecular patterns (PAMPs). Typical PAMPs include lipopolysaccharide (LPS) and peptidoglycan from the cell walls of gram-negative and gram-positive bacteria, respectively, bacterial flagellin, and microbial DNA and RNA.

Toll-Like Receptors

TLRs are the principal PAMP receptors in the innate immune system and obtained their unusual name because of their similarity to the receptor

Toll (German for "funky" or "cool") in the fruit fly *Drosophila*. Toll was initially a cause for scientific excitement when it was found to be responsible for dorsoventral body patterning in *Drosophila*, but, in addition, was later shown to form part of the fly's immune defense against fungal infections. This phylogenetically ancient system of pathogen detection is highly conserved in evolution, with similar receptors occurring not only in humans and invertebrates, but also in plants such as tobacco.

Eleven different TLRs have been identified in mammals. The first to have its involvement in pathogen recognition demonstrated, and the most studied, is TLR4, which responds to the most powerful stimulant of innate immune responses, gram-negative bacterial endotoxin (LPS). This was established through study of two strains of mice that fail to mount a septic response to large doses of endotoxin and that were shown to have a loss-of-function mutation in the gene for TLR4. Subsequently, other TLRs and their ligands have been identified; these are summarized in Table 2.1. Some TLRs are able to respond to microbial ligands on their own, but, in many cases, the response depends on the interaction of several different molecules at the cell surface. TLR dimers are required for signaling through TLR4 (homodimers of two TLR4 molecules) and TLR2 (heterodimers with either TLR1 or TLR6, with the combination determining the ligand specificity of the receptor complex). In addition, LPS signaling through TLR4 requires the interaction of several other molecules at the receptor complex; LPS is delivered to the receptor by soluble LPS binding protein (LBP), and effective receptor activation requires the presence of at least two additional molecules, CD14 and MD2.

There is differential subcellular localization of individual TLRs. TLR2 and TLR4 are expressed on the cell surface, where they are most likely to encounter material from microbial cell walls. TLR3 and TLR9 are located within endosomes, where they are most likely to encounter their ligands in the lytic products of phagocytosed microorganisms.

At first, the triggering of immune responses by a relatively small range of receptors and ligands might seem crude. However, most microorganisms present more than one TLR ligand, therefore, it

TABLE 2.1. TLRs and their known ligands

TLR	Ligands
TLR1 (heterodimer with TLR2)	Triacylated lipopeptides, lipomannans from *Mycobacterium tuberculosis*
TLR2 (often dimer with TLR2 or 6)	Lipoproteins, peptidoglycans, lipoteichoic acids, yeast zymosan
TLR3	Double-stranded RNA
TLR4 (homodimer plus CD14 and MD2)	LPS, heat shock proteins, pneumolysin, respiratory syncytial virus coat proteins, heparan sulphate fragments, fibrinogen peptides
TLR5	Flagellin
TLR6 (heterodimer with TLR2)	Diacylated lipopeptides
TLR7	Responds to synthetic nucleosides and imidazoquinoline antivirals; native ligand is thought to be single-stranded RNA in endosomes
TLR8	Same as for TLR7
TLR9	Bacterial DNA—unmethylated CpG motifs
TLR10	Ligand unknown but TLR10 expressed in lung and B lymphocytes
TLR11	Uropathogenic bacteria in mice; absent in humans

is likely that microbes with differing patterns of molecular motifs can cause differential activation of a number of TLRs, allowing differential responses to various classes of pathogen.

TLR Signaling

Understanding of the signaling pathway through which TLR ligation leads to activation of a cell and secretion of inflammatory mediators has advanced considerably in the last few years. The end product of intracellular signal transduction is activation of transcription factors, which translocate to the nucleus and modulate transcription of target genes. The principal transcription factor in inflammation is nuclear factor κB (NFκB), which up-regulates transcription of genes for inflammatory mediators such as TNFα, ILs, and cyclooxygenase (COX)-2. Other transcription factors under TLR regulation induce proapoptotic, antiapoptotic, and even anti-inflammatory gene transcription, although how the differential effects of these pathways are modulated is not yet well understood.

Apart from TLR3, all TLRs signal down a common pathway accessed via the adaptor molecule, myeloid differentiation factor (MyD)-88. The various signaling intermediates have been

identified and are likely to be the targets of future immunomodulatory therapies in sepsis and inflammatory disease, and therefore, are summarized in Figure 2.1.

MyD88 recruits a kinase, IL-1 receptor-associated kinase (IRAK)-4, and facilitates its phosphorylation of IRAK-1. IRAK-1 then associates with TNF receptor-associated factor (TRAF)-6 to activate the transforming growth factor-β-activating kinase (TAK)-1/TAK1 binding protein (TAB) complex, which, in turn, enhances the activity of the inhibitor of NFκB (IκB) kinase (IKK) complex. NFκB is held inactive in the cytoplasm by its inhibitor, IκB. The IKK complex phosphorylates IκB, leading to its degradation and the release of free NFκB, which can translocate to the nucleus. There, NFκB undergoes phosphorylation and associates with other transcription regulators to activate inflammatory gene transcription.

TLR3 and TLR4 can also access a separate MyD88-independent pathway to inflammatory gene transcription using the adaptor molecules toll/IL-1 receptor domain-containing adaptor-inducing interferon-β (TRIF) and TRIF-related adaptor molecule (TRAM). This pathway leads to a slower activation of NFκB and also to transcription of genes for the type 1 interferons via a different transcription factor, interferon regulatory factor (IRF)-3.

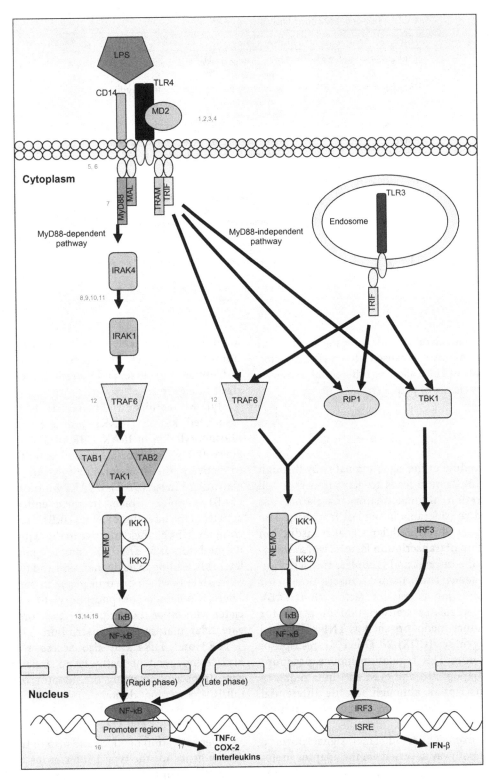

FIGURE 2.1. Initiation of inflammation through TLRs. Intracellular signaling through other TLRs uses the MyD88-dependent pathway with small variations. Numbers in blue indicate the site of action of the regulators of TLR signaling summarized in Table 2.2. MAL, MyD88 adaptor-like; NEMO, NFκB essential modulator; RIP1, receptor-interacting protein 1; TBK1, TRAF-family member-associated NFκB activator-binding kinase 1; ISRE, IFN-stimulated response element.

TABLE 2.2. Known regulators of TLR signaling[a]

	Regulator	Action
1	Soluble TLRs	Bind ligand and prevent interaction with cell-surface TLR2 and TLR4; soluble TLR4 can also bind MD2 making it unavailable to the LPS receptor complex
2	Triad3A	Ubiquitylates TLRs, marking them for degradation
3	RP105	Inhibits ligand binding to TLR4
4	MD2B	Inactive variant of MD2
5	Single immunoglobulin IL-1-related receptor (SIGIRR)	Sequesters adaptor molecules MyD88 and MAL
6	ST2	Sequesters adaptor molecules
7	MyD88s (short)	Splice variant of MyD88, antagonizes MyD88
8	Suppressor of cytokine signaling (SOCS)-1	Inhibits IRAK
9	Phosphatidylinositol 3 kinase (PI3K)	Mechanism still unknown
10	IRAK-M	Inhibits IRAK phosphorylation
11	Toll-interacting protein (TOLLIP)	Inhibits phosphorylation and facilitates degradation of IRAK
12	A20	Inhibits TRAF6
13	IκB	Activated NFκB increases transcription of its own inhibitor, proving negative feedback control
14	TNF-related apoptosis-inducing ligand receptor (TRAILR)	Stabilizes IκB, so more NFκB is retained in the cytoplasm
15	Nucleotide-binding oligomerization domain (NOD)-2	Activated by bacterial muramyl dipeptide and suppresses NFκB; defective in Crohn's disease
16	Inhibitors of transcription	Inhibitory forms of NFκB and IκBξ block NFκB DNA binding; chromatin remodeling (changes in methylation and acetylation) alters rate of transcription
17	Reduced stability of messenger RNA	Reduces synthesis of inflammatory compounds

[a]Sites of action are marked with blue numbers in Figure 2.1.

Regulation of TLR Signaling

TLR activation can trigger a rapid and vigorous inflammatory response, therefore, it is not surprising that TLR signaling is subject to regulation at multiple levels. Some regulatory molecules are constitutively expressed in tissues and plasma, whereas others are induced by activation of the TLR signaling pathway and, thus, provide negative feedback regulation. There is negative feedback within the signaling pathway itself with the gene for the inhibitor IκB being under direct control of a NFκB-binding promoter sequence, thus, NFκB activation results in increased IκB concentration and subsequent down-modulation of the NFκB effect. Known regulators of TLR signaling are summarized in Table 2.2 and their site of action marked on Figure 2.1.

Endotoxin Tolerance

Repeated observations have demonstrated that the natural history of an episode of sepsis or SIRS consists of the initial inflammatory phase of vigorous innate immune responses, and then a period of relative immune suppression in which the individual is at increased susceptibility to further infections. These secondary infections tend not to elicit as vigorous an inflammatory response as the initial infection and can insidiously become widespread. This is paralleled by the responses of isolated monocytes, which, after an initial stimulation with LPS, show diminished proinflammatory cytokine responses to repeat stimulation. This phenomenon of "endotoxin tolerance" has also been demonstrated with other TLR responses, with previous exposure to a TLR ligand producing diminished responses to the same TLR ligand (termed "homotolerance") and, in some cases, to other TLR ligands ("heterotolerance").

The mechanism of TLR tolerance is still being investigated and some of the regulators of TLR signaling mentioned above have been implicated in its etiology. Other mechanisms may involve down-regulation of surface TLRs or nuclear events that suppress the transcription of proinflammatory genes. Recent work has suggested that endotoxin tolerance is not simply an all-or-nothing "off switch" for inflammation, but rather a state of immune "reprogramming"—a switch to more anti-inflammatory cytokine profiles with modulation of LPS sensitivity, so that markedly increased doses can still induce an inflammatory response. It is becoming clear that the surrounding cytokine milieu can modulate the effect of tolerance with, for instance, interferon-γ restoring LPS sensitivity in some systems.

Endotoxin tolerance may have developed as a protective mechanism to avoid death from the cytokine storm associated with severe sepsis but, in the age of intensive care, it puts the postseptic patient in danger of later infective complications. Treatments designed to reverse tolerance and "reboot" the innate immune response might give hope of improving survival after sepsis. However, these will have to be developed with caution because endotoxin tolerance may be significant in situations other than sepsis. Organ systems, such as the gut and liver, that are exposed to tonic levels of TLR ligands from commensal microbes may rely on the tolerance mechanisms to physiologically elevate their threshold for activation and prevent unwanted inflammation.

Clinical Directions

The coming years will see increasing relevance of TLR signaling to clinical practice. Polymorphisms of genes for TLRs and components of the signaling pathway have already been shown to influence severity of sepsis and susceptibility to invasive bacterial disease. Increased understanding of the "inflammatory" and "tolerant" phases of the septic response may help novel anti-inflammatory and immunostimulatory therapies to be used appropriately.

References

1. Annane D, Bellissant E, Cavaillon JM. Septic shock. *Lancet* 2005;365:63–78.
2. Opal SM, Huber CE. Bench-to-bedside review: Toll-like receptors and their role in septic shock. *Crit Care* 2002;6:125–136.
3. Takeda K, Akira S. Toll-like receptors in innate immunity. *Int Immunol* 2005;17:1–14.
4. Liew FY, Xu D, Brint EK, O'Neill LAJ. Negative regulation of toll-like receptor-mediated immune responses. *Nat Rev Immunol* 2005;5:446–458.
5. Sabroe I, Read RC, Whyte MKB, et al. Toll-like receptors in health and disease: complex questions remain. *J Immunol* 2003;171:1630–1635.

3
Metabolic and Endocrine Changes in Sepsis and the Catabolic State

Steven G. Ball

Sepsis and the Catabolic State

Sepsis and critical illness constitute severe physical stress, associated with a characteristic physiological response. Historically, the metabolic and endocrine components of this response have been considered part of a uniform and (should the situation go on) sometimes persisting physiological adaptation. More recently, this model has been challenged on two fronts. Firstly, the metabolic and endocrine responses to sepsis and critical illness are not uniform and persistent. Rather, acute and chronic phases (where they occur) are associated with responses that are both distinct and discordant. Secondly, the interpretation that these responses are somehow adaptive has been rechallenged. Specific patterns of metabolic and endocrine response have prognostic value. Furthermore, recent clinical studies have demonstrated the benefit of intervention targeting specific metabolic and endocrine endpoints. Both of these observations challenge the simple assumption that these responses are a positive design feature. Natural selection could have resulted in metabolic adaptation to acute severe illness built, in part, on the physiological responses to other stressors. However, the same process could not have produced metabolic adaptation to persisting critical illness—an entity that has only existed following developments in organ support and critical care units. What is adaptive in the short term may not be thereafter. Rather, these responses may go on to form part of the pathophysiological cascade that characterizes multiple organ failure.

This review will outline some of the key metabolic and endocrine responses to sepsis and critical illness; highlighting the mediators and sequelae of these responses where they are known; the basis for intervention; and the outcome data after intervention where available.

Metabolic-Nutritional Responses to Sepsis and Critical Illness

Although undernutrition often accompanies prolonged illness, prolonged critical illness and sepsis are associated with a metabolic response that differs from that of uncomplicated undernutrition. Protein conservation does not occur. Rather, there is increased resting energy expenditure and proteolysis (especially in skeletal muscle) leading to a profound negative nitrogen balance, with urinary losses of 11 to 14 g in 24 hours—a negative balance that cannot be reversed simply by nutritional support. This is accompanied by increased gluconeogenesis, hepatic glucose output, and lipolysis. In addition, there are marked changes in trace elements, with a decrease in plasma zinc and iron levels with an increase in copper levels consequent to increased hepatic ceruloplasmin production—an acute phase reactant. These changes can be prolonged if the underlying problem persists. The endocrine system has a role in both the initiation and maintenance of these metabolic responses. Moreover, the relationship is reciprocal; metabolic changes in turn impacting on endocrine function. This tautology, coupled with the uncertainty regarding whether either or both responses contribute to adverse outcome, has fueled debate regarding the role of nutritional/metabolic and endocrine intervention in this complex area of pathophysiology.

The Insulin-Glucose Axis in Sepsis and Critical Illness

Hyperglycemia is one of the characteristic metabolic responses to critical illness, brought about through the combined effects of counter-regulatory hormones such as cortisol and growth hormone (GH), the sympathetic nervous system and inflammatory cytokines. The initial phase of critical illness is characterized by α-adrenergic inhibition of β-cell insulin release, despite concurrent hyperglycemia. In addition, there is both elevated hepatic glucose output and impaired muscle glucose uptake, features of insulin resistance. Subsequently, insulin production returns to normal or is elevated through the action of islet β-adrenergic receptors. Insulin resistance persists, however, and can result in continued glucose intolerance.

In those with prolonged critical illness, strict maintenance of normoglycemia with intensive insulin therapy can reduce mortality by half (Van den Berghe et al., 2001). The precise mechanism whereby maintenance of blood glucose produces these outcomes remains uncertain. The intervention is associated with a number of potential positive mediators; reduction in dialysis-dependent acute renal failure; enhancement of impaired immune function; and normalization of dyslipidemia.

Endocrine Responses to Sepsis and Critical Illness

The Hypothalamo-Pituitary-Adrenal Axis

The hypothalamo-pituitary-adrenal (HPA) axis has a pivotal role in the homeostatic response to severe illness, cortisol having a key role in maintaining vascular physiology and mediating acute metabolic responses. Acute critical illness,

including sepsis, results in marked, centrally driven increases in cortisol production by the adrenal cortex. Corticotropin-releasing factor (CRF) and adrenocorticotropin (ACTH) drive is augmented by the action of vasopressin and inflammatory cytokines. The resultant hypercortisolism stimulates gluconeogenesis and inhibits anabolic processes; augments circulating volume and both the pressor and inotropic effects of catecholamines; and dampens potentially damaging exuberant immune responses.

Persistent critical illness is associated with continued hypercortisolemia. However, ACTH levels are low, suggesting that adrenal glucocorticoid production is driven by alternative agents. Both natriuretic peptides and substance P have been proposed as mediators of ACTH-independent adrenal steroid production in this context. In addition, there is a peripherally driven shift in adrenal steroidogenesis away from mineralocorticoid and androgen production in favor of glucocorticoids.

ACTH-independent cortisol production may fail in some patients. Furthermore, cortisol production in patients with persistent critical illness can be increased by exogenous ACTH, indicating that cortisol production may be submaximal in this setting despite the physiological state and the fact that ACTH levels may be limiting when other trophic stimuli fail. In recent years, the concept of relative adrenal insufficiency has been developed. This term describes apparent inappropriate cortisol levels in patients with persistent critical illness/sepsis syndrome and, by inference, serves to identify a group of patients who may benefit from treatment with exogenous glucocorticoids. This area remains controversial. The definition and use of the term lack precision and consistency, and there remain problems defining an appropriate normal reference range in this context.

Specific features of the HPA axis have prognostic value in sepsis. Patients with elevated baseline cortisol levels (>935 nmol/L) and a relatively small increment following a standard 250 μg ACTH stimulation test (<250 nmol/L) have a mortality of 80% (Annane et al., 2000). A large multicenter randomized trial has shown that, in patients with septic shock who demonstrate an increment of plasma cortisol below this level on ACTH stimulation testing, treatment with exogenous glucocorticoids at doses thought to mimic levels found in acute pathological stress both reduces vasopressor dependency and improves 28-day survival (Annane et al., 2002). These data have led support to the concept of relative adrenal insufficiency as an important parameter influencing outcome in critical illness. However, there are a number of issues that complicate the interpretation of these data.

Only some 5% of total plasma cortisol circulates in the free form, the rest being bound in one of two plasma protein components. Some 80 to 90% of cortisol circulates bound to high-affinity cortisol-binding globulin (CBG), whereas some 5 to 10% is bound loosely to a low-affinity site on albumin. Both CBG and albumin levels commonly fall in sepsis and critical illness. This may lead to a situation in which total plasma cortisol is an underestimate of the hormone that is biologically active. It has been suggested that low cortisol binding proteins (BPs) rather than reduction in biologically active free cortisol underlies the observation of apparent reduction in cortisol in critical illness (Hamrahian et al., 2004).

Diagnostic criteria based on increments of cortisol in response to ACTH stimulation lack validity in the context of acute illness. The increment in plasma cortisol in response to ACTH is inversely correlated with baseline cortisol; the higher the baseline cortisol, the lower the incremental response to ACTH—in critical illness, the patient may be maximally stimulated at baseline (May and Carey, 1998). In the intervention study of Annane et al. (2002), the patients with the lowest increment in cortisol after ACTH stimulation also had lower baseline cortisol levels. This suggests that the group may have contained some patients with true cortisol insufficiency.

The anesthetic agent, etomidate, inhibits the activity of the steroidogenic enzymes cholesterol (P450) side-chain cleavage enzyme (P450 SCC) and 11-β hydroxylase (Wagner et al., 1984). In some but not all patients, the degree of inhibition may be sufficient to affect the outcome. Anesthetic induction with etomidate was only established as an exclusion criteria midway through the study. A recent observational study of children with meningococcal sepsis found that both total and free cortisol was significantly lower in those

children that did not survive the illness than in those that did survive (den Brinker et al., 2005). These differences did not reflect polymorphisms in steroidogenic enzyme activity. Importantly, however, there was a striking independent association between adverse outcome and both interleukin (IL)-6 levels and the use of etomidate as an induction agent for anesthesia.

Subsequent studies have raised questions regarding the broad application of intervention with glucocorticoids based on incremental responses to ACTH stimulation. Widemer et al. (2005) demonstrated that 40% of patients recovering from coronary artery bypass grafting did not achieve an increment in plasma cortisol of 250 nmol/L with ACTH stimulation at the time of extubation. Such a response was not associated with a negative outcome. Using an elevated base-line cortisol (>550 nmol/L) and a response to ACTH of less than 250 nmol/L as indicative of relative adrenal insufficiency, Pizarro et al. (2005) found no correlation between relative adrenal insufficiency and increased mortality in pediatric septic shock. Together, these data would suggest that incremental response to ACTH should not be used as the sole indicator of relative adrenal insufficiency in critical illness (Marik, 2004; Dickstein, 2005).

In patients with septic shock, 240 mg exogenous hydrocortisone every 24 hours administered independently of endogenous adrenal function generates plasma cortisol levels of approximately 3300 nmol/L, much higher than those found in acute severe stress (Keh et al., 2003). This raises the possibility that intervention may highlight benefit of pharmacological treatment rather than physiological replacement. Indeed, a recent meta-analysis has demonstrated benefit of low-dose, moderate duration (5–7 d) glucocorticoid treatment in patients with septic shock independent of adrenal function (Minneci et al., 2004). This is in contrast to historical studies that used higher doses and for shorter periods. Previous approaches may simply have resulted in the dominance of established adverse effects of glucocorticoids over potential benefits.

The Hypothalamo-Pituitary-Thyroid Axis

The thyroid axis exhibits a biphasic response to critical illness. Within a few hours of onset, free triiodothyronine (FT)-3 levels fall, whereas FT4, thyroid-stimulating hormone (TSH), and rT3 (the inactive product of T4 inner-ring deiodination) levels increase. These changes reflect reduced peripheral T4 to T3 conversion and reduced degradation of rT3. Tumor necrosis factor (TNF)-α, IL-1, and IL-6 are likely mediators of this response, because they can produce similar changes in experimental models. Importantly, however, cytokine antagonism does not prevent the development of this classic low-T3 syndrome in human disease models. Increased degradation of thyroxine-binding globulin (TBG) and inhibition of thyroid hormone binding to carrier proteins through the action of bilirubin and free fatty acids also contribute to reduced T3 levels at a tissue level. These changes at the very onset of critical illness/sepsis are followed by a normalization of TSH and FT4, but persistent subnormal levels of FT3. The nighttime TSH surge can be lost.

Should critical illness persist, additional changes in the thyroid axis can develop during several weeks, with a reduction in thyroid hormone production. Both FT4 and FT3 levels fall into the subnormal range. TSH values fall into the subnormal or low-normal range, and pulsatile TSH secretion is lost. Postmortem data indicate that this is secondary to decreased hypothalamic TRH production, consistent with the proposal that the persistent reduced FT3 state is centrally driven. This resetting of the thyroid neuroendocrine axis is accompanied by additional changes in peripheral thyroid hormone metabolism, with down-regulation of hepatic type 1 deiodinase (D1) and up-regulation of type 3 deiodinase (D3). These changes further accentuate the alteration in active/inactive hormone balance; the FT3/rT3 ratio (Peeters et al., 2005).

The GH, Insulin-Like Growth Factor, Insulin-Like Growth Factor-BP Axis

In normal conditions, the anterior pituitary produces the majority of GH as a series of nighttime pulses. Both pulse amplitude and pulse frequency contribute to integrated GH levels; interpulse trough GH levels being very low. The acute phase of critical illness is characterized by elevated integrated GH production, with an increase in pulse amplitude and a marked elevation in interpulse trough GH values. Concentrations of the GH-BP (derived from the GH receptor) and the GH-dependent proteins, insulin-like growth factor (IGF)-1 and IGF-BP3 fall. These data are consistent with the development of peripheral GH resistance.

Persistent critical illness is accompanied by additional changes in the GH axis. There is reduced pulsatile GH release, with predominant reductions in pulse amplitude. Interpulse trough GH levels are reduced in comparison with the acute phase, but remain elevated with respect to the normal state. GH-BP (and by inference GH receptor expression) levels return to normal. However, GH-dependent IGF-1 and IGF-BP3 levels remain low. These data suggest the development of a relative GH insufficiency with normalization of peripheral GH sensitivity, and raise the possibility that this combination contributes to the progressive negative nitrogen balance and wasting that further characterizes this clinical situation.

A large multicenter study of high-dose GH treatment in adults with persisting critical illness demonstrated a worsening of morbidity and a doubling of mortality (Takala et al., 1999). Current data would suggest that the use of high-dose GH in the context of normal peripheral GH sensitivity—the likely context of the study, would not be appropriate.

Hepatic IGF-BP1, a small circulating IGF-BP, is regulated by metabolic status. Production is inversely related to hepatic substrate availability and inhibited by insulin. IGF-BP1 levels are higher in nonsurvivors of critical illness in comparison with survivors. To what extent these reflect a partial causal relationship or an epiphenomenon reflecting substrate deficiency or relative insulin insufficiency remains to be established (Van den Berghe, 2000). These data serve to highlight the interrelationship between nutrition, metabolism, and endocrinology in this complex area.

The Hypothalamo-Pituitary-Gonadal Axis

Many catabolic states, including sepsis, starvation, and critical illness, are associated with biochemical hypogonadism. However, as with several of the other endocrine axes, the pituitary-gonadal axis responses in the acute and chronic settings are somewhat different.

Immediately after an acute event, such as myocardial infarction or surgery, testosterone production is suppressed whereas luteinizing hormone (LH) levels are high—indicative of hyperacute suppression of Leydig cell function. The mechanism of this suppression is not clear. Inflammatory cytokines (IL-1 and IL-2) are key candidates. Prolonged critical illness is associated with the development of progressive hypogonadotrophic hypogonadism. LH pulse frequency may increase, but pulse amplitude is decreased. In the male, testosterone levels may become very low, whereas estradiol concentrations remain in the normal range—indicative of enhanced peripheral aromatization. TNF-α stimulates activity of the aromatase complex (Zhao et al., 1996). Aromatization of adrenal androgens may serve to maintain sex hormone levels in the female to some extent.

Implications for Clinical Practice

The endocrine response to stress is an attractive therapeutic target in severe sepsis. The surviving sepsis guidelines currently recommend both tight insulin control and cortisol replacement therapy. However, the stress response is both complex and time dependent, calling into question the likely efficacy of simple replacement schedules. The maintenance of near normoglycemia with intensive insulin therapy in patients with critical illness is supported by one randomized controlled trial (RCT), and others are about to report. A multicenter RCT examining steroid replacement, the Corticosteroid Therapy of Septic Shock (CORTICUS) trial, is also nearing completion. However, in the meantime, the lack of well designed and powered clinical trials of other endocrine intervention means that any additional approaches should be seen as experimental and considered as part of a study protocol.

Major metabolic/endocrine changes in early sepsis

- Hyperglycemia and insulin resistance
- Increased cortisol release
- Increased TSH and free T4 levels

- Reduced free T3 and tissue T3 levels
- Increased GH release and peripheral GH resistance

Major metabolic/endocrine changes in persisting critical illness

- Hyperglycemia and insulin resistance
- ACTH-independent cortisol production
- Reduced TSH and reduced thyroid hormone production
- Reduced pulsatile GH release and normalization of GH sensitivity

Problems in interpretation of the cortisol response in sepsis

- Appropriate/optimal response unknown
- Most circulating cortisol is protein bound
- Few studies have measured free cortisol in sepsis
- ACTH response depends on baseline cortisol levels, which are variably elevated in sepsis
- Some drugs (e.g., etomidate) depress cortisol release
- Even "low dose" hydrocortisone produces supraphysiological circulating levels of cortisol

References

Annane D, Sebille V, Charpentierc, Bollaert PE, Francois B, Korach JM, Capellier G, Cohen Y, Azoulay E, Troche G, Chaumet-Riffaut P, Bellissant E. Effect of treatment with low doses of hydrocortisone and fludrocortisone on mortality in patients with septic shock. *JAMA* 2002;288:862–871.

Annane D, Sebille V, Toche G, Raphael JC, Gajdos P, Bellisant E. A 3-level prognostic classification in septic shock based on cortisol levels and cortisol response to corticotropin. *JAMA* 2000;283:1038–1045.

Den Brinker M, Joosten KFM, Liem O, de Jong FH, Hop WCJ, Hazelet JA, van Dijk M, Hokken-Koelega ACS. Adrenal insufficiency in meningococcal sepsis: bioavailable cortisol levels and impact of interleukin-6 levels and intubation with etomidate on adrenal function and mortality. *J Clin Endocrinol Metab* 2005;90:5110–5117.

Dickstein G. Editorial: on the term 'relative adrenal insufficiency'—or what do we really measure with adrenal stimulation tests? *J Clin Endocrinol Metab* 2005;90:4973–4974.

Hamrahian AH, Oseni TS, Arafah BM. Measurements of serum free cortisol in critically ill patients. *N Engl J Med* 2004;350:1629–1638.

Keh D, Boehnke T, Weber-Cartens S, Schulz C, Ahlers O, Bercker S, Volk HD, Doecke WD, Falke KJ, Gerlach H. Immunologic and haemodynamic effects of 'low-dose' hydrocortisone in septic shock. *Am J Respir Crit Care Med* 2003;167:512–520.

Marik PE. Unraveling the mystery of adrenal insufficiency in the critically ill. *Crit Care Med* 2004;32: 596–597.

May ME, Carey RM. Rapid adrenocorticotropic test in practice. *Am J Med* 1985;79:679–684.

Minneci PC, Deans KJ, Banks SM, Eichaker PQ, Natanson C. Meta-analysis: the effect of steroids on survival and shock during sepsis depends on the dose. *Ann Intern Med* 2004;141:47–56.

Pizarro CF, Troster EJ, Damiani D, Carcillo JA. Absolute and relative adrenal insufficiency in children with septic shock. *Crit Care Med* 2005;33: 855–859.

Peeters RP, Wouters PJ, van Toor H, Kaptein E, Visser TJ, Van den Berghe. Serum 3,3′,5′-triiodothyronine (rT3) and 3,5,3′-triiodothyronine / rT3 are prognostic markers in critically ill patients and are associated with post-mortem deiodinase activities. *J Clin Endocrinol Metab* 2005;90: 4559–4565.

Takala J, Ruokonen E, Webster NR, Nielson MS, Zandsra DF, Vundelinckx G, Hinds CJ. Increased mortality associated with growth hormone treatment in critically ill adults. *N Engl J Med* 1999;341:785–792.

Van den Berghe G. Novel insights into the Neuroendocrinology of critical illness. *Eur J Endocrinol* 2000;143:1–13.

Van den Berghe G, Wouters P, Weekers F, Verwaest C, Bruyninckx F, Schetz M, Vlasselaers D, Ferdinande P, Lauwers P, Bouillon R. Intensive insulin therapy in critically ill patients. *N Engl J Med* 2001;345: 1357–1367.

Wagner RL, White PF, Kan PB, Rosenthal MH, Feldman D. Inhibition of adrenal steroidogenesis by the anesthetic etomidate. *N Engl J Med* 1984;310:1415–1421.

Widemer IE, Puder JJ, Konig C, Pargger H, Zerkowski HR, Girard J, Muller B. Cortisol response in relation to the severity of stress and illness. *J Clin Endocrinol Metab* 2005;90:4579–4586.

Zhao Y, Nichols JE, Valdez R, Mendelson CR, Simpson ER. Tumor necrosis factor-α stimulates aromatase gene expression in human adipose stromal cells through use of an activating protein-1 binding site upstream of promoter 1.4. *Mol Endocrinol* 1996;10: 1350–1357.

4
Hematological and Coagulation Changes in Sepsis

Tina T. Biss and J. Wallace-Jonathan

Introduction

The septic patient poses a particular challenge to the clinician and hematologist because of the numerous ways in which the blood count and coagulation profile may be altered and the dynamic nature of these changes. This chapter summarizes the major ways in which sepsis can alter the blood and outlines areas of controversy in management.

Hematological Changes in Sepsis

- **Neutrophil leucocytosis** is the usual response to bacterial sepsis in an adult, often with a "left shift," whereby immature myeloid cells, normally confined to the bone marrow, are released into the peripheral blood. Neutrophils and precursors show increased or "toxic" granulation and vacuolation (see Figure 4.1). Very rarely, in severe sepsis, phagocytosed organisms are observed within the neutrophils. A very high, left-shifted count is termed a "leukemoid reaction," although the total white cell count rarely exceeds 60×10^9 cells/L and the proportion of blasts is relatively low.[1]
- In contrast, some patients with sepsis, particularly the very young and the elderly, present with **neutropenia**. This is caused by the adherence of neutrophils to activated endothelial cells lining blood vessel walls (termed "margination") before migration into the surrounding tissue where they may be consumed at sites of infection. This phase is usually short-lived and recovery occurs after 24 to 48 hours. Preexisting neutropenia, which may be drug-related or caused by underlying marrow disease, may, of course, lead to severe infection and this distinction can be difficult in the acute phase. A bone marrow biopsy may be necessary if neutropenia persists. Patients with liver disease and hypersplenism may be neutropenic but usually have an adequate neutrophil response to infection. Nutritional deficiencies and the drugs used to treat infection may contribute to prolongation of the neutropenic phase. In patients receiving chemotherapy, the rapid development of neutropenia within 12 to 24 hours may be the first sign of sepsis.

Treatment of neutropenia with recombinant human granulocyte colony-stimulating factor (G-CSF) or granulocyte-macrophage colony-stimulating factor (GM-CSF) may increase the peripheral blood neutrophil count, but there is no data from randomized controlled trials to support use in acute sepsis in adults without a preexisting bone marrow abnormality. Combined data from two small studies in neonates[2,3] showed a survival benefit of colony-stimulating factors (CSF) when used in systemic infection accompanied by neutropenia. A recent Cochrane review[4] concludes that this warrants further investigation. Although there is currently insufficient evidence to support the routine use of CSF in neonatal practice, as prophylaxis or to treat established infection, it may be helpful in selected cases.

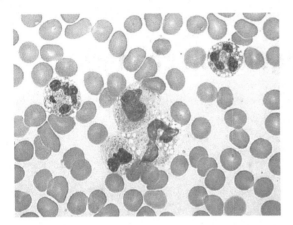

FIGURE 4.1. Blood film showing "toxic" neutrophils, with coarse cytoplasmic granules and vacuolation of the cytoplasm. Two-band forms and a myelocyte are also seen, and the patient is thrombocytopenic. (Kindly photographed by Dr. M.M. Reid)

- **Thrombocytopenia** in sepsis is common and is often multifactorial. The following should be considered:
  - Preexisting problems
  1. Liver disease and hypersplenism
  2. Alcohol excess
  3. Folate deficiency
  These may reduce platelet count independently, or may coexist in the same patient who is then also predisposed to infection.
  4. Marrow suppression caused by immunosuppressive agents or chemotherapy
  5. Immune thrombocytopenia
  - Immediate effects of infection
  1. DIC
  2. Platelet consumption secondary to infection without overt coagulopathy
  - Later effects of infection and treatment
  1. Severe acidosis and multiorgan dysfunction
  2. Platelet destruction within hemodialysis/hemofiltration circuits
  3. Folate deficiency developing during treatment
  4. Drugs that suppress marrow platelet production, e.g.:
     - Linezolid
     - Chloramphenicol
  5. Drugs that cause immune thrombocytopenia, e.g.:
     - Heparin
     - Teicoplanin

  6. Hemophagocytic syndrome
  - Artifact
  1. Thrombocytopenia caused by ethylenediamine tetra-acetic acid (EDTA)-induced platelet aggregation
  2. Difficult venipuncture
- **Thrombocytopenia and outcome:** Recent studies have shown that thrombocytopenia (platelet count $<150 \times 10^9$ cells/L) has a prevalence of up to 44% in adult medical intensive care unit (ICU) patients,[5,6] the commonest cause being sepsis with or without overt DIC.[5] Thrombocytopenia in this setting is associated with increased mortality, particularly in those with a low platelet nadir ($<100 \times 10^9$ cells/L) or a fall in platelet count of greater than 30 to 50% from admission.[5,6] After correction for contributory factors, an independent association between nadir platelet count and ICU death was observed in one of these studies,[6] and a fall in platelet count of greater than 30% from admission, but not thrombocytopenia per se, was an independent predictor of mortality in the other.[5] Thrombocytopenic patients have a higher bleeding rate and greater requirement for transfusion of blood products.[5] After an initial fall in platelet count, recovery from thrombocytopenia within 7 days predicts a better prognosis than thrombocytopenia without platelet count recovery.[7,8] It is, therefore, recommended that ICU patients undergo serial measurements of platelet count and that the underlying cause of thrombocytopenia be actively sought and managed.
- **Thrombocytosis** is part of the acute phase response. The platelet count may exceed 1000×10^9 cells/L and usually peaks at 10 to 20 days after the onset of infection.[9] The presence of uniformly small platelets on the blood film, in addition to a normal platelet count before the septic event, makes "reactive thrombocytosis" more likely than primary thrombocytosis/myeloproliferative disorder.[1] A recent study of 176 intensive therapy unit (ITU) trauma patients[10] showed that 20.4% developed thrombocytosis (platelet count $>600 \times 10^9$ cells/L), the most common cause being infection. Unlike in the myeloproliferative disorders, thrombocytosis was not associated with increased risk of thromboembolism and predicted a better outcome than predicted by severity of illness

score. Platelet count normalized a median of 35 days (range, 17–73 d) after ICU admission. No specific treatment for reactive thrombocytosis is required; thromboprophylaxis should be used according to usual routine. Platelet count should be monitored after the acute event has settled to ensure that it returns to normal.

- **Anemia** in the septic patient is usually caused by a combination of causes, including frequent venipuncture, loss of red cells in dialysis circuits, bleeding, folate deficiency, hemolysis, and anemia of chronic disease.

Coagulation Changes in Sepsis

Sepsis frequently causes hemostatic abnormalities. These range from insignificant laboratory findings to overt DIC, which is associated with increased mortality and increased incidence of organ failure.[11]

Disseminated Intravascular Coagulation

Definition

Tissue factor (TF; also known as thromboplastin) is the key initiator protein and thrombin is the key effector protein of coagulation. The key coagulant, anticoagulant, fibrinolytic, and antifibrinolytic pathways are shown in Figure 4.2. DIC occurs when the carefully balanced system of procoagulant and anticoagulant forces is overwhelmed by a massive systemic procoagulant signal. Thus, coagulation, normally a localized and appropriate process, becomes systemic, resulting in intravascular fibrin formation. Procoagulant and anticoagulant proteins, platelets, and fibrinogen are all consumed, leading to widespread hemostatic failure. This produces a spectrum of clinical features caused by thrombosis or bleeding or both.[12]

Pathogenesis of Sepsis-Related DIC

The activation of coagulation in sepsis is primarily mediated by the up-regulation of TF expression, predominantly on monocytes and endothelial cells, in response to increased levels of tumor necrosis factor and other cytokines occurring in sepsis. Inflammatory damage to endothelial cells results in increased permeability and exposure of underlying tissue that bears TF, collagen, and other coagulation activators. Activation of the coagulation system can also occur on the surface of the bacterium itself.[13]

Levels of the procoagulant factors and the anticoagulant factors, protein C and antithrombin, are markedly reduced because of consumption, although circulating activated protein C, a small fraction of total protein C, may actually be increased (protein C requires activation by exposure to thrombomodulin, in the presence of protein S, before it is able to inactivate factors Va and VIIIa; see Figure 4.2). Depletion of protein C and antithrombin III encourage the sustained thrombin generation that defines sepsis-related DIC. Plasma levels of plasminogen-activating inhibitor (PAI)-1 are elevated, resulting in impaired fibrin degradation caused by inhibition of tissue plasminogen activator (tPA).[13] The significance of this change is underlined by the findings discussed below.

Some inherited tendencies, such as protein C deficiency, particular polymorphisms of the PAI-1 gene,[14] or acquired deficiencies, such as liver disease, may make a patient particularly liable to develop DIC when faced with a septic challenge. The gene for PAI-1 has a common polymorphism (4G/5G) in the promoter sequence that, in the homozygous form (occurring in 27% of the UK

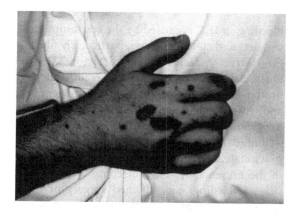

FIGURE 4.2. Purpura fulminans in a patient with meningococcal septicemia and DIC. (Reproduced by kind permission from Dr. C.H. Toh)

population), leads to significantly higher plasma levels of PAI-1. In several independent studies in children with meningococcal sepsis, it has been shown that this polymorphism is associated with a significant increase in incidence of serious thrombosis and death. Studies of the PAI-1 polymorphism and outcome in adults with sepsis have not been reported.

Diagnosis

DIC is a clinicopathological diagnosis, taking both clinical and laboratory features into account.[15]

Clinical Features

DIC can present anywhere on the spectrum of bleeding to thrombosis, although thrombosis is the dominant clinical problem in DIC associated with sepsis:

- Fibrin deposition can result in localized thrombus formation or, more commonly, widespread microvascular thrombosis causing organ dysfunction. The most dramatic form of this is referred to as "purpura fulminans," which results from fibrin deposition in the dermal vasculature, leading to hemorrhagic tissue necrosis (see Figure 4.2). This is most often a feature of bacterial sepsis, particularly meningococcal or pneumococcal infection, but may be a consequence of viral infection, especially in the presence of hereditary protein C deficiency.
- Consumption of coagulation factors and platelets may result in primary failure of hemostasis, termed "consumption coagulopathy." Systemic activation of plasminogen results in inappropriate fibrinolysis after initial hemostasis has occurred. This failure of both formation and stabilization of clot presents with oozing and bleeding from numerous sites (wounds, venipuncture sites, indwelling line/drain sites, mucous membranes, and gastrointestinal and genitourinary tracts).

Laboratory Diagnosis

The characteristic laboratory findings in DIC are shown in Table 4.1.

TABLE 4.1. Characteristic laboratory findings in DIC

Hematological parameter	Characteristic finding in DIC
Routinely measured	
Platelets	Reduced; usually to $<100 \times 10^9$ cells/L or falling rapidly
Prothrombin time	Prolonged
Activated partial thromboplastin time	Prolonged
Fibrinogen	Reduced; or falling rapidly
Fibrinogen degradation products/D-dimers	Increased; often markedly increased
Blood film	Red cell fragmentation; particularly in cancer-related DIC with thrombotic sequelae
Second-line investigations/research tools	
Protein C/protein S/antithrombin III	Reduced
PAI-1	Increased

Source: Adapted from Taylor et al., 2001 [15].

Management

The mainstay of management of sepsis-related DIC is treatment of the underlying infection. However, the use of antibiotics in the presence of bacterial infection does not always result in resolution of the coagulopathy. Because DIC covers a spectrum of clinical sequelae from asymptomatic to severe thrombotic or bleeding manifestations, management of the coagulopathy can range from monitoring only to aggressive replacement therapy with blood products (see below). Directed therapies have been used with varying success and are detailed below.

Blood Product Support in DIC

Bleeding in DIC is caused by the consumption of platelets and clotting factors, particularly fibrinogen, factor V, factor VIII, and factor XIII (the fibrin-stabilizing protein).[13] This is illustrated by prolonged clotting times, hypofibrinogenemia, and thrombocytopenia on laboratory testing. Replacement of clotting factors by transfusion of a combination of fresh-frozen plasma (FFP), cryoprecipitate, and platelet concentrates is advisable in the presence of active bleeding or significant risk of bleeding, e.g., planned surgery or other invasive procedures. In the absence of active bleeding or other indication, there is no evidence to support the prophylactic transfusion of FFP or cryoprecipitate, regardless of the severity of the laboratory abnormalities.[16,17] Indications for transfusion of blood products are as follows:

- **Transfusion of platelets:** In the absence of bleeding, platelet count should be maintained at greater than 20×10^9 cells/L. If there is bleeding or an invasive procedure is planned, the threshold is raised to 50×10^9 cells/L. This is adequate for most invasive procedures (including lumbar puncture, liver biopsy, transbronchial biopsy, insertion of indwelling lines, and laparotomy). An exception is surgery involving a critical site, such as the brain, spinal cord, or eye, for which a platelet count of greater than 100×10^9 cells/L should be maintained.[16] Transfusion of one platelet concentrate to an adult should raise the platelet count by at least 20×10^9 cells/L, but, in the presence of a consumptive coagulopathy, bleeding, or splenomegaly, a larger dose may be required. It is difficult to give specific advice and we recommend that the platelet count be measured after platelet transfusion to ensure that an adequate response has occurred. Young children should be transfused with 15 mL/kg of platelet concentrate, whereas older children may receive an adult dose.[16]
- **Transfusion of FFP:** If there is active bleeding or an invasive procedure is planned, a prolonged prothrombin or activated partial thromboplastin time is an indication for transfusion of FFP. The recommended dose of 10 to 15 mL/kg usually approximates to 4 U for an adult of 70-kg body weight.[17] The response depends on the degree of coagulopathy and the rapidity of coagulation factor consumption. Assuming that plasma volume is approximately 3.5 L and that this returns to 3.5 L after infusion of 1 L of plasma, the theoretical increase in factor levels would be approximately 30%. However, some factors, such as factor IX, have considerable extravascular distribution, reducing the effective rise, and others, particularly factor VII, have a short half life ($t_{1/2} = 6$ h). Again, the recommended approach is to monitor response both clinically and by frequent blood sampling.
- **Transfusion of cryoprecipitate:** In the presence of active bleeding, transfusion of cryoprecipitate is recommended to maintain a fibrinogen level of greater than 1.0 g/L. A dose of 10 mL/kg usually equates to 10 U for an adult of 70-kg body weight.[17] These are pooled into a single bag before being issued by the blood bank. Each unit of cryoprecipitate, made by concentrating the less soluble proteins (fibrinogen, factor VIII, von Willebrand factor, and vitronectin) into a smaller volume, contains approximately half the quantity of these proteins present in the original unit of FFP. Thus, 10 U of cryoprecipitate contains similar amounts of fibrinogen and factor VIII to 5 U of FFP. A dose of cryoprecipitate will increase the fibrinogen by 0.5 to 1 g/L.

It is essential that frequent testing of platelet count and coagulation parameters occurs, at least once daily and after every episode of blood product transfusion. Measurement of D-dimer levels does not influence the use of blood products but, as a measure of activity of the coagulation system, monitoring a trend can be helpful in determining whether the coagulopathy is improving or worsening.

Infective risks of transfusion are now quite insignificant compared with the risk of death caused by inadequate treatment of severe sepsis. Of noninfective risks, severe and transfusion-related acute lung injury (TRALI) caused by donor HLA antibodies is seen with approximately 1 in 8000 U of unselected FFP.[18] A retrospective study from the Mayo clinic suggests that acute lung injury (ALI) is strongly associated with transfusion of FFP, and that this may be a milder unrecognized form of TRALI interacting with other pulmonary insults.[19] In England, FFP is preferentially from male donors, and early evidence suggests that this has significantly reduced the risk of TRALI. Platelets and red cells from female donors also contain some plasma, and ALI related to transfusion of these blood components remains possible.

Evidence of a significant immunomodulatory effect related to transfusion, that might increase mortality from infection, is conflicting and doubtful. It is possible that non-leucodepleted components may have harmful effects, but all UK blood components are now leucodepleted. Increased mortality after a liberal red cell transfusion regime (of non-leucodepleted components) was observed in the Transfusion Requirements in Critical Care (TRICC) study.[20] This seemed to be associated with an increased incidence of lung injury and cardiac problems, which may have been caused, in part, by TRALI. Following the TRICC study, most intensivists use a hemoglobin (Hb) level of 7 g/dL as a threshold for red cell transfusion, but a higher level (8–9 g/dL) may be appropriate for patients with known vascular disease or with bleeding in whom a hematocrit of greater than 30% (Hb, ~10 g/dL) can improve hemostasis.

Anticoagulant Drugs in Sepsis-Related DIC

Assuming that the organ dysfunction observed in sepsis-related DIC is a thrombotic phenomenon, a number of drugs with anticoagulant properties have been used in an attempt to reduce end-organ damage. Only activated protein C has been shown to improve patient outcome.

Heparin

It has been suggested that treatment with heparin, in an attempt to reduce coagulation activation and promote fibrinolysis, would result in reduction of microvascular occlusion and subsequent organ damage.[21] Beneficial responses have been shown in animal models[22] but no randomized trials in clinical practice have been published. One concern is the potential for heparin to aggravate bleeding. Although anecdotal reports describe its successful use in patients with chronic compensated DIC and thrombotic manifestations, the use of heparin in septic DIC cannot be routinely recommended.

Antithrombin III

Reduced levels of natural anticoagulants are a feature of DIC, and a significant reduction in antithrombin III level at the onset of septic shock correlates with a poor prognosis.[23] Although in vitro studies have shown antithrombin III to have an anti-inflammatory effect in addition to an anticoagulant effect,[24] a randomized controlled trial of antithrombin III (30,000 U administered as a 6000 IU intravenous loading dose followed by continuous intravenous infusion of 6000 IU/d for 4 d) versus placebo in 2314 patients with sepsis and DIC showed rapid resolution of DIC but no survival advantage in those treated with antithrombin III.[25] When used in conjunction with heparin, bleeding rates were increased.[25]

Protein C Concentrate in DIC

Reduction in protein C activation is a pathological finding in children with severe meningococcal sepsis. A placebo-controlled, dose-finding study randomized 40 children with meningococcal septic shock and purpura fulminans to receive recombinant human protein C concentrate at varying doses.[26] This study showed a dose-dependant increase in activated protein C levels in the treatment group, but the numbers were too small to demonstrate a difference in outcome.

Activated Protein C

In addition to anticoagulant properties, activated protein C also has an anti-inflammatory and an antiapoptotic role.[27] This protects the endothelium from damage by proinflammatory cytokines and thrombin deposition, thus, reducing organ damage. The Recombinant Human Activated Protein C [Xigris] Worldwide Evaluation in Severe Sepsis (PROWESS) study[28] randomized 1690 adult patients with severe sepsis and organ failure to receive either recombinant human activated protein C (drotrecogin alfa) or placebo. All-cause mortality at 28 days was significantly reduced in the treatment group (24.7% vs. 30.8%; $P = 0.005$) at the cost of a nonsignificant increase in serious bleeding complications (3.5% vs. 2%; $P = 0.06$). The beneficial effect of drotrecogin alfa (activated) was greatest in those with highest risk of mortality, in both the presence and absence of DIC,[29] whereas the risk of bleeding exceeded the potential benefit in those with an APACHE II score of less than 25 or single-organ failure.[30] Activated protein C is currently indicated in adults at high risk of death from sepsis (APACHE II score of $\geq$25, sepsis-induced multiple organ failure, septic shock, or sepsis-induced acute respiratory distress syndrome) with or without DIC. Contraindications to therapy include severe thrombocytopenia (platelets $<30 \times 10^9$ cells/L) and a high risk of bleeding, such as that caused by recent trauma, recent hemorrhagic stroke, or gastrointestinal ulceration.[31] Concurrent administration of heparin at greater than prophylactic doses should be avoided. The half-life of drotrecogin alfa is short, allowing invasive procedures or surgery to be performed 2 hours after the infusion is stopped. The recommended dose of drotrecogin alfa (activated) is 24 µg/kg body weight/h as a 96-hour continuous intravenous infusion.[31] The recent Extended Evaluation of Recombinant Human Activated Protein C (ENHANCE) study, an international single-arm trial of drotrecogin alfa in 2378 adults with severe sepsis, has supported the efficacy seen in PROWESS but showed higher rates of serious bleeding events both during (3.6% vs. 2.4%) and after (3.2% vs. 1.2%) infusion when compared with the PROWESS study treated group. Earlier treatment in this study was shown to improve outcome.[32] However, a trial in children with severe sepsis has been stopped early and did not show benefit. These studies raise some doubts regarding the safety of drotrecogin alfa and further trials in high risk subgroups of adults have been proposed.

tPA in DIC

Meningococcal disease causes septic shock and associated DIC. In severe cases, widespread vascular thrombosis occurs, resulting in purpura fulminans and irreversible ischaemia of digits and limbs. Fibrinolysis is impaired by raised levels of PAI-1, and variation in the PAI-1 gene has been shown to influence severity of DIC and outcome.[14] Thrombolysis with recombinant tPA (rt-PA) was studied retrospectively in 62 infants and children with severe meningococcal sepsis treated for predicted amputations and/or refractory shock.[33] The median dose was 0.3 mg/kg/h for a median duration of 9 hours. It is possible that rt-PA prevented amputation, because 12 of the 50 patients treated for expected amputation did not undergo amputation. However, mortality was high, at 47%, and intracranial hemorrhage occurred in 5 of 62 patients (3 of whom died). Conclusions cannot be drawn from this study because of the lack of an untreated control group.

TF Pathway Inhibitor in DIC

TF is a major coagulation activator in sepsis. The use of TF pathway inhibitor (TFPI) in treatment of sepsis was studied in 1955 patients in a placebo-controlled multicenter trial.[34] Recombinant TFPI (tifacogin) at a dose of 0.025 mg/kg/h intravenously for 96 hours was used in the treatment group. There was no difference in all-cause mortality at 28 days, but an increased incidence of serious bleeding occurred in the tifacogin group (6.5% vs. 4.8%).

Plasmapheresis in DIC

Observational studies have suggested a benefit from plasmapheresis, correction of coagulation abnormalities being clearly demonstrated.[35] However, blinded trials of plasmapheresis are not practical and it may never be possible to prove a survival benefit.

Other Coagulation Changes in Sepsis

- A prolonged prothrombin time can be caused by reduced levels of vitamin K-dependant factors as a result of sepsis-related liver failure or malnutrition. Parenteral feeds are particularly low in lipid content and, therefore, vitamin K, which is a lipid-soluble vitamin. Adult intensive care patients should routinely receive 10 mg of vitamin K orally/intravenously three times per week and children should receive 0.3 mg/kg.[16] Prophylactic transfusion of FFP to correct prolonged clotting times is not recommended in the absence of DIC and bleeding.[16]
- A low fibrinogen level may be caused by liver failure but may also relate to generalized malnourishment. Fibrinogen replacement by transfusion of cryoprecipitate is not recommended in the absence of bleeding.[16]
- A raised fibrinogen level, in addition to raised levels of factor VIII, von Willebrand factor, protein C, and protein S, are part of the acute phase response.

Conclusion

DIC and sepsis remain a major challenge for the intensivist. Numerous treatments have failed to live up to their early promise. It may be that treatment targeted to those with a higher risk genetic profile for thrombotic problems will improve outcome in the future.

References

1. Bain BJ. Blood cells: a practical guide. 3rd ed. 2002, Malden, MA, Oxford, Blackwell.

2. Bilgin K, Yaramis A, Haspolat K, Tas MA, Gunbey S, Derman O. A randomized trial of granulocyte-macrophage colony-stimulating factor in neonates with sepsis and neutropenia. *Pediatrics* 2001;108: 1383–1384.

3. Kocherlakota P, La Gamma EF. Human granulocyte colony-stimulating factor may improve outcome attributable to neonatal sepsis complicated by neutropenia. *Pediatrics* 1997;100:E6.

4. Carr R, Modi N, Dore C. G-CSF and GM-CSF for treating or preventing neonatal infections. *Cochrane Database Syst Rev* 2003;3:CD003066.

5. Strauss R, Wehler M, Mehler K, Kreutzer D, Koebnick C, Hahn EG. Thrombocytopenia in patients in the medical intensive care unit: bleeding prevalence, transfusion requirements, and outcome. *Crit Care Med* 2002;30:1765–1771.

6. Vanderschueren S, De Weerdt A, Malbrain M, Vankersschaever D, Frans E, Wilmer A, Bobbaers H. Thrombocytopenia and prognosis in intensive care. *Crit Care Med* 2000;28:1871–1876.

7. Nijsten MW, ten Duis HJ, Zijlstra JG, Porte RJ, Zwalveling JH, Paling JC, The TH. Blunted rise in platelet count in critically ill patients is associated with worse outcome. *Crit Care Med* 2000;28:3843–3846.

8. Acka S, Haji-Michael P, de Mendonca A, Suter P, Levi M, Vincent JL. Time course of platelet counts in critically ill patients. *Crit Care Med* 2002;30: 753–756.

9. Griesshammer M, Bangerter M, Sauer T, Wennauer R, Bergmann L, Heimpel H. Aetiology and clinical significance of thrombocytosis: analysis of 732 patients with an elevated platelet count. *J Intern Med* 1999;245:295–300.

10. Valade N, Decailliot F, Rebufat Y, Heurtematte Y, Duvaldestin P, Stephan F. Thrombocytosis after trauma: incidence, aetiology, and clinical significance. *Br J Anaesth* 2005;94:18–23.

11. Gando S, Nanzaki S, Kemmotsu O. Disseminated intravascular coagulation and sustained systemic inflammatory response syndrome predict organ dysfunctions after trauma: application of clinical decision analysis. *Ann Surg* 1999;229:121–127.

12. Toh CH, Dennis M. Disseminated intravascular coagulation: old disease, new hope. *Br Med J* 2003;327:974–977.

13. Dempfle CE. Coagulopathy of sepsis. *Thromb Haemost* 2004;91:213–224.

14. Haralambous E, Hibberd ML, Hermans PW, Ninis N, Nadel S, Levin M. Role of functional plasminogen-activator-inhibitor-1 4G/5G promoter polymorphism in susceptibility, severity, and outcome of meningococcal disease in Caucasian children. *Crit Care Med* 2003;31:2788–2793.

15. Taylor FB, Toh CH, Hoots WK, Wada H, Levi M. Towards definition, clinical and laboratory criteria and a scoring system for disseminated intravascular coagulation. *Thromb Haemost* 2001;86: 1327–1330.

16. British Committee for Standards in Haematology, Blood Transfusion Task Force. Guidelines for the use of platelet transfusions. *Br J Haematol* 2003; 122:10–23.

17. British Committee for Standards in Haematology, Blood Transfusion Task Force. Guidelines for the use of fresh-frozen plasma, cryoprecipitate and cryosupernatant. *Br J Haematol* 2004;126: 11–28.

18. Wallis JP. Transfusion related acute lung injury (TRALI)-under-diagnosed and under-reported. *Br J Anaesth* 2003;90:573–576.

19. Gajic O, Rana R, Mendez JL, Rickman OB, Lymp JF, Hubmayr RD, Moore SB. Acute lung injury after blood transfusion in mechanically ventilated patients. *Transfusion* 2004;44:1468–1474.

20. Hebert PC, Wells G, Blajchmann MA, Marshall J, Martin C, Pagliarello G, Tweeddale M, Schweitzer I, Yetisir E. A multicenter, randomized, controlled clinical trial of transfusion requirements in critical care. *N Engl J Med* 1999;340:409–417.

21. Tanaka T, Tsujinaka T, Kambayashi J, Higashiyama M, Yokota M, Sakon M, Mori T. The effect of heparin on multiple organ failure and disseminated intravascular coagulation in a sepsis model. *Throm Res* 1990;60:321–330.

22. Slofstra SH, Van't Veer C, Buurman WA, Reitsma PH, ten Cate H, Spek CA. Low molecular weight heparin attenuates multiple organ failure in a murine model of disseminated intravascular coagulation. *Crit Care Med* 2005;33:1365–1370.

23. Mesters RM, Mannucci PM, Coppola R, Keller T, Ostermann H, Kienast J. Factor VIIa and antithrombin III activity during severe sepsis and septic shock in neutropenic patients. *Blood* 1996;88:881–886.

24. Oelschlager C, Romisch J, Staubitz A, Stauss H, Leithauser B, Tillmanns H, Holschermann H. Antithrombin III inhibits nuclear factor kappa B activation in human monocytes and vascular endothelial cells. *Blood* 2002;99:4015–4020.

25. Warren BL, Eid A, Singer P, Pillay SS, Carl P, Novak I, Chalupa P, Atherstone A, Penzes I, Kubler A, Knaub S, Keinecke HO, Heinrichs H, Schindel F, Juers M, Bone RC, Opal SM. Caring for the critically ill patient. High-dose antithrombin III in severe sepsis: a randomized controlled trial. *JAMA* 2001; 286:1869–1878.

26. de Kleijn ED, de Groot R, Hack CE, Mulder PG, Engl W, Moritz B, Joosten KF, Hazelzet JA. Activation of protein C following infusion of protein C concentrate in children with severe meningococcal sepsis and purpura fulminans: A randomized, double-blinded, placebo-controlled, dose-finding study. *Crit Care Med* 2003;31:1839–1847.

27. Joyce DE, Gelbert L, Ciaccia A, DeHoff B, Grinnell BW. Gene expression profile of antithrombotic protein C defines new mechanisms modulating inflammation and apoptosis. *J Biol Chem* 2001; 276:11199–11203.

28. Bernard GR, Vincent JL, Laterre PF, LaRosa SP, Dhainaut JF, Lopez-Rodriguez A, Steingrub JS, Garber GE, Helterbrand JD, Ely EW, Fisher CJ Jr. Efficacy and safety of recombinant human activated protein C for severe sepsis. *N Engl J Med* 2001;344:699–709.

29. Dhainaut JF, Yan SB, Joyce DE, Pettila V, Basson B, Brandt JT, Sundin DP, Levi M. Treatment effects of drotrecogin alfa (activated) in patients with severe sepsis with or without overt disseminated intravascular coagulation. *J Thomb Haemost* 2004;2: 1924–1933.

30. Abraham E, Laterre PF, Garg R, Levy H, Talwar D, Trzaskoma BL, Francois B, Guy JS, Bruckmann M, Rea-Neto A, Rossaint R, Perrotin D, Sablotzki A, Arkins N, Utterback BG, Macias WL. Drotrecogin alfa (activated) for adults with severe sepsis and a low risk of death. *N Engl J Med* 2005;353: 1332–1341.

31. Fourrier F. Recombinant human activated protein C in the treatment of severe sepsis: An evidence-based review. *Crit Care Med* 2004;32[Suppl.]: S534–S541.

32. Vincent J-L, Bernard GR, Beale R, Doig C, Putensen C, Dhainaut J-F, Artigas A, Fumagalli R, Macias W, Wright T, Wong K, Sundin D, Turlo MA, Janes J. Drotrecogin alfa (activated) treatment in severe sepsis from the global open-label trial ENHANCE: Further evidence for survival and safety and implications for early treatment. *Crit Care Med* 2005; 33:2266–2277.

33. Zenz W, Zoehrer B, Levin M, Fanconi S, Hatzis TD, Knight G, Mullner M, Faust SN. Use of recombinant tissue plasminogen activator in children with meningococcal purpura fulminans: A retrospective study. *Crit Care Med* 2004;32:1777–1780.

34. Abraham E, Reinhart K, Opal S, Demever I, Doig C, Rodriguez AL, Beale R, Svoboda P, Laterre PF, Simon S, Light B, Spagen H, Stone J, Seibert A, Peckelsen C, De Deyne C, Postier R, Pettila V, Artigas A, Percell SR, Shu V, Zwingelstein S, Tobias J, Poole L, Stolzenbach JC, Creasey AA. Efficacy and safety of tifacogin (recombinant tissue factor pathway inhibitor) in severe sepsis: a randomized controlled trial. *JAMA* 2003;290:256–258.

35. Stegmayr BG, Banga R, Berggre L, Norda R, Rydvall A, Vikefors T. Plasma exchange as rescue therapy in multiple organ failure including acute renal failure. *Crit Care Med* 2003;31:1730–1736.

5
The Genetics of Sepsis and Inflammation

Martin F. Clark

Classic population genetic techniques suggest a strong genetic component to both the risk of developing sepsis and the subsequent outcome in terms of survival. For example, adoption studies indicate an approximate six-fold increase in the risk of premature death from infection if an adoptees biological parent died from sepsis before the age of 50 years.

New techniques have been developed to investigate the genetic influence on complex diseases. The discovery of common variations in human DNA, termed genetic polymorphisms, has led to the publication of a large number of genetic association studies in human sepsis.

What Are Genetic Polymorphisms and How Can They Be Used to Investigate the Genetics of Sepsis?

Genetic polymorphisms are stable variations in the DNA sequence. By convention, each polymorphism must occur in at least 1% of the population. Most polymorphisms involve a single base pair substitution, known as single nucleotide polymorphisms (SNPs), e.g., adenine changing to cytosine (denoted as either A→C or 1→2, where 1 is the commonest form of the polymorphism and 2 is the rarer form; (Figure 5.1). Polymorphisms are also numbered to denote where in the gene they occur, for example, tumor necrosis factor (TNF)-α–308 represents a polymorphism 308 base pairs upstream of the transcription initiation site, a positive number denotes a polymorphism downstream of this site. More complex variations, such as insertions and deletions, are also recognized. Polymorphisms occur on average every 200 to 1000 DNA base pairs, although they can occur more frequently (noncoding regions) or less frequently (protein-coding DNA) in some areas of the genome. Because each cell contains 6 billion DNA base pairs, this equates to several million DNA base pair differences and roughly 100,000 amino acid differences between any two individuals. The detection of these polymorphisms is relatively straightforward, using standard restriction fragment length polymorphism (RLFP) and other molecular biology techniques.

Polymorphisms may have no effect (they are neither transcribed nor affect transcription), may alter gene transcription (polymorphism occurs in gene promoter region), or may change the structure of proteins coded by affected genes, thus, potentially altering protein function. An artificial distinction is made between polymorphisms and mutations, mutations being polymorphisms with sufficient impact to cause a disease state, e.g., cystic fibrosis, whereas polymorphisms are said to merely alter disease risk.

In genetic association studies, candidate gene(s) with single or multiple polymorphisms are chosen. Case-control, cohort, or a combination study design is used. In the case–control design, the frequency of the polymorphism is determined in both patients with sepsis and controls, and the risk of death or other adverse events are compared with the frequency of the polymorphism to establish a possible association. In cohort studies,

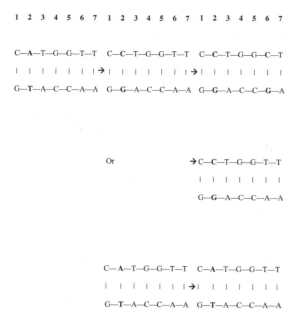

FIGURE 5.1. SNP and LD. The first polymorphism involves an A→C substitution (this could also be written as T→G) at position 2. If a second polymorphism then occurs near this mutated segment (T→C at position 6), the new mutation is in LD. That is, the presence of a C at position 6 means that position 2 must be a C, whereas a T at position 6 means that position 2 could be either A or C.

the frequency of the polymorphism in survivor and nonsurvivor groups is usually compared.

It has become apparent that there are major potential problems inherent in genetic association studies:

1. All genetic association studies are highly susceptible to producing false-positive associations (Type I error). If every polymorphism in the genome is tested for an association with sepsis, using a P value of 0.05 (5% chance of Type I error), then approximately 300,000 positive associations would occur purely by chance. If we assume that 50 polymorphisms actually affect sepsis, then the chance of a positive result (at a significance of $P = 0.05$) being a true positive is 50 in 300,000; that is, 0.017%. It has been estimated that to reach a 95% probability of no false-positives results on a genome-wide association study, a P value of 5×10^{-8} is required.

Clearly, research should only target polymorphisms occurring in likely candidate genes, but most authorities suggest that the odds of any one targeted polymorphism actually having a true effect

are still approximately 100:1 to 1000:1. Thus, false-positive results are still highly likely, especially when publication bias is taken into account. Therefore, it is not surprising that a comprehensive review of more than 600 genetic association studies, assessing 166 polymorphisms, found only 6 polymorphisms that demonstrated consistent replication of effect. In light of this, the journal *Nature Genetics* refuses to publish genetic association studies, unless investigators perform a second confirmatory study for any association claimed.

2. Linkage disequilibrium (LD) is another problem. LD occurs when two polymorphic forms (alleles) at separate loci on the chromosome occur together more frequently than predicted by chance (Figure 5.1). LD is complete when the presence of an allele at one site is 100% predictive of the allele at the second polymorphic site. Thus, a polymorphism may be linked to a disease, not because it has any effect, but because it is in LD with a polymorphism that does. In addition, polymorphisms can exist in LD with polymorphisms that have an opposite or additive effect to their action. Thus, studies among different ethnic populations may produce varying results depending on the degree of LD that exists between polymorphisms in any given population.

3. The in vivo effects of polymorphisms cannot be predicted from in vitro data, nor can their effects be generalized across patient groups, because a different model may produce different results. Sepsis and systemic inflammatory response syndrome (SIRS) are nonhomogeneous, complex clinical syndromes involving multiple molecular systems, which interact and regulate each other. The effect of any polymorphism may be minimized by compensatory mechanisms that keep the immune response balanced. For a polymorphism to overcome this, and produce a measurable effect, it must first significantly alter the expression or function of a gene, and second, the molecule coded by this gene must be sufficiently powerful biologically, to alter the whole system response. A good example is a chemokine receptor polymorphism which inhibits human immunodeficiency virus (HIV) entering cells, and thus prevents HIV infection in exposed individuals.

4. Association studies in sepsis often have major methodological flaws in control group selection, genetic assay technique, study blinding,

statistical interpretation, study replication, study size, and power. Many studies are underpowered, enrolling fewer than 100 septic patients, whereas approximately 2000 patients has been suggested as a suitable study size. Inadequate sample size increases the chance of false-negative results and further increases the large false-positive association risk. Bayesian analysis suggests that many studies reporting a positive association between a genetic polymorphism and sepsis are likely to be false-positive associations.

Genetic Association Studies in Sepsis (Figure 5.2)

Coagulation Polymorphisms

These are the most promising polymorphisms assessed to date.

Plasminogen activator inhibitor (PAI) inhibits both tissue and urine plasminogen activators, which may lead to a procoagulant state, disseminated intravascular coagulation, and increased microvascular thrombus. A commonly occurring polymorphism of the PAI-1 gene consists of a single base pair insertion (5G) or deletion (4G) of a guanine molecule at position −675 of the PAI gene. The 4G form cannot bind a transcription repressor

protein and is associated with a 25% increase in transcription and plasma levels of PAI.

All four studies assessing this polymorphism reported an increased mortality associated with the 4G form, although only three studies reached significance and one of these studies had major methodological flaws. The fourth study was an underpowered cohort containing only 29 septic patients. The 4G/4G genotype has also been associated with increased vascular complications (amputation, skin graft, or plastic surgery referral).

The factor V Leiden mutation results in an arginine-to-glycine substitution at amino acid 506 of the factor V molecule, which confers partial resistance to inactivation by activated protein C (APC). This increases the risk of macrovascular thrombosis but, crucially, may decrease the risk of microvascular thrombosis because it raises the level of endogenous APC via increased thrombin generation. In the Recombinant Human Activated Protein C [Xigris] Worldwide Evaluation in Severe Sepsis (PROWESS) trial, mortality of heterozygous factor V carriers was 13.9% versus 27.9% for noncarriers ($P = 0.013$).

Mannose Binding Lectin

Mannose-binding lectin (MBL) binds microbial surface carbohydrates, initiating opsonophagocy-

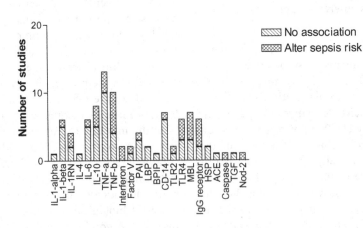

FIGURE 5.2. Major genetic association case–control studies examining the risk of developing sepsis according to the class of polymorphism. The results of many of these studies are contradictory and it is difficult to draw firm conclusions from the results. LBP, lipopolysaccharide binding protein; BPIP, bactericidal/permeability-increasing protein; HSP, heat-shock protein; ACE, angiotensin-converting enzyme; TGF, transforming growth factor; Nod, nucleotide oligomerization domain. (Intensive Care Medicine, Volume 32, Issue 11, 2006: 1706–1712. A Systematic Review of the Quality of Genetic Association Studies. In Human Sepsis by Clark, M and Baudouin, S. Reprinted with kind permission from Springer Science and Busniess Media.)

tosis. The MBL gene has five polymorphisms, any of which decreases serum MBL levels from 100% in those carrying no polymorphisms (denoted A/A) to 10% in those with one polymorphism (A/O) and to 1% in those with two polymorphisms (O/O), i.e., it is almost a gene knockout. Despite this, these polymorphisms have no effect on mortality and have only been shown to have a significant effect on sepsis development in four of seven studies. This may be because MBL is not absolutely vital in the response to infection and other pathways can compensate for a decrease in its activity.

CD14

CD14 binds to a wide range of molecules but especially lipopolysaccharide. High soluble CD14 levels have been associated with increased mortality in gram-negative septic shock. A CD14–159T→C polymorphism that increases circulating CD14 levels has been investigated in sepsis, but studies have not demonstrated an effect from this polymorphism.

Immunoglobulin Gamma

Immunoglobulin gamma (Fc) receptors facilitate complement activation, phagocytosis, and antibody dependent cellular toxicity. FcγRIIa is the only immunoglobulin receptor that binds IgG 2 (the major receptor for encapsulated bacteria). It has a polymorphism coding for amino acid 131, which results in a histidine-to-arginine substitution. The homozygous histidine form (H/H) has greater phagocytic activity against Haemophilus, Staphylococci, and Neisseria species compared with the heterozygous (H/R) or arginine homozygous (R/R) forms.

Studies on this polymorphism in meningitis have reported contradictory results. Two poor-quality studies have reported an approximately two-fold risk for developing pneumococcal sepsis in R/R genotype patients versus H/R or R/R. Thus, an association with pneumococcal sepsis is possible but not proven.

Toll-Like Receptors

Toll-like receptors (TLR) bind pathogenic organisms and their components, subsequently increasing expression of transcription factors for proinflammatory cytokines. TLR4 binds lipopolysaccharide and plays a major role in the response to gram-negative infections. Two polymorphisms of TLR4 have been demonstrated in humans; both result in amino acid substitutions but do not affect TLR4 activity in vitro. Based on the available studies, TLR polymorphisms are unlikely to have an effect on the development or outcome from sepsis.

Proinflammatory Cytokines

Polymorphisms in interleukin (IL)-1α, IL-1β, IL-6, TNFα, and TNFβ have been investigated.

The evidence for IL-1 is poor, however the largest study to date (1106 meningitis patients) did report an association between the IL-1β-511 low IL-1 secretor phenotype, IL-1β-511(1/+), and improved survival.

IL-6 induces an acute phase response. It has stimulatory effects on B lymphocytes, T lymphocytes, and macrophages. A polymorphism involving a G→C substitution exists at position −174. The C allele is associated with lower IL-6 production. However, the evidence for an association between this polymorphism and sepsis is weak.

TNFα is a major proinflammatory cytokine, and administration reproduces the deleterious effects of endotoxemia and bacteraemia.[28] High circulating TNFα levels correlate with poorer outcome in septic states.[29] Inhibition of TNFα, by antibodies to TNFα or soluble TNF receptors, administered before the infective insult, prevents sepsis developing in animal models. However, trials of TNFα antibodies in human septic patients failed to improve outcome. The TNFα-308G→A polymorphism has been extensively studied in sepsis, with the A phenotype increasing TNFα production in some studies but not others. The commonest genotyping technique for this polymorphism has been demonstrated to be flawed and misclassifies the genotype in 9% of patients assayed. There is no strong evidence for a clinical effect from this polymorphism.

The TNFβ gene lies near the TNFα gene and has a polymorphism that exists as either a guanine (TNF-B1) or adenine (TNF-B2), which corresponds to an asparagine-to-threonine substitution at amino acid 26 of the TNFβ molecule. The TNFβ2/2 form is associated with increased secretion of TNFα and has been shown to be in LD with the TNFα-308G allele (low secretion phenotype). The TNFβ2/2 polymorphism is associated with development of severe sepsis in several small pre-intensive care unit (ICU) cohort studies, but a definitive association cannot be supported yet. It is unlikely to influence mortality in patients with sepsis.

Anti-Inflammatory Cytokines

IL-1 receptor antagonist (IL-1RN) competitively binds to the IL-1 receptor, preventing IL-1 binding. Production is controlled by the IL-1RN gene, which is polymorphic. The IL-1RN2 form is associated with increased IL-1RN production in vitro but not reproducibly in vivo. The evidence for an effect in sepsis is poor.

Il-10 is a major counter inflammatory cytokine that inhibits synthesis of proinflammatory cytokines. High plasma levels have been shown to correlate with the development of infection in trauma patients, organ failure in cardiac surgery patients,[4] and negative outcome from sepsis. The IL-10-1082G→A polymorphism in the G form increases stimulated IL-10 release from ex vivo blood[3] of infected patients but not in vivo. Studies to date are generally small and do not support an association with sepsis for this polymorphism.

Combinations of Polymorphisms

Because the effect from any single polymorphism tends to be minimized by the interaction of regulatory factors, there is increasing interest in looking at haplotype clades. These are evolutionary groups of several polymorphisms that exist in LD with one another, in a gene or segment of DNA, and are inherited as a unit. It is proposed that the effect of these clades may be more biologically powerful and, thus, more likely to produce a clinically detectable effect. This approach has recently been used to claim an association between IL-6 clades and increased mortality in SIRS patients.

Similarly, some studies have assessed the effect of combinations of polymorphisms that have a synergistic effect on the immune response. Patients with the low IL-1β secreting polymorphism, IL-1β-511(1/+), and the high IL-1RN secreting polymorphism, IL-1RN2/+, have decreased IL-1 activity. In one study of meningitis, patients with both polymorphisms had increased mortality. However, such subgroup analyses are at high risk of Type I error and no study has assessed polymorphism combinations as an a priori finding.

Clinical Uses for Polymorphisms

At present, polymorphisms have no clinical application and remain still very much a research tool. It has been suggested that polymorphisms could be used to target therapy to a patient's genetic make-up, i.e., anti-TNFα therapy for high TNFα producers. However, there is, to date, no definitive evidence for an effect from any single polymorphism, especially TNFα polymorphisms, in sepsis. Furthermore, we cannot extrapolate the in vitro effect of polymorphisms to in vivo and we do not yet understand the pathophysiology of sepsis sufficiently to allow targeted interventions based on natural human genetic variation.

The Future

Larger, better-designed studies are being conducted (for example, the UK GAINS study and the European GENOSEPT study), which will give results that are more definitive. The study of haplotype clades, or other polymorphism combinations is also promising. It is well established that individual genetic variation has an important impact on the risk and outcome of human sepsis. Identifying the key genes that produce these differences is proving to be difficult despite huge advances in molecular biology. However, the exponential growth in our understanding of human genetics continues and it is likely that within the next decade the important genetic determinants of sepsis will be discovered.

Suggested Reading

Buckland PR. Genetic association studies of alcoholism—problems with the candidate gene approach. Alcohol and alcoholism. 2001;36(2):99–103.

Brookes AJ. The essence of SNP's. Gene 1999;234: 177–186.

Glossary of Terms

Terms	Definitions	Terms	Definitions
Allele	One or more alternate forms of a DNA sequence	Genotype	The genetic identity of an individual that does not show as outward characteristics
Bayesian analysis	Calculates a false-positive report probability (FPRP); this estimates the likelihood of a study reporting a false-positive association. The factors that determine the size of the FPRP are 1) the previous probability of a true association of the tested genetic variant with a disease; 2) the observed P value; and 3) the statistical power to detect the odds ratio of the alternative hypothesis at the given P value. A high FPRP (>0.5) could, therefore, be a consequence of any combination of a low previous probability, a low power, or a high P value	Heterozygous	Two different alleles at a given locus
		Homozygous	Two identical alleles at a given locus
		Insertion	A type of chromosomal abnormality in which a DNA sequence is inserted into a gene, disrupting the normal structure and function of that gene
		Intron	A noncoding segment of the gene that is spliced out of the mature RNA product
		Gene Knockout	A genetically engineered organism that carries one or more genes in its chromosomes that have been made inoperative
		Locus	A specific location on a chromosome. Alternatively, the position on a chromosome at which the gene for a particular trait resides
Chromosome	The self-replicating genetic structure of cells containing the cellular DNA that bears in its nucleotide sequence the linear array of genes. Chromosomes are normally found in pairs; human beings typically have 23 pairs of chromosomes.	LD	Linkage disequilibrium; two alleles (or a trait and an allele) at different loci that occur together within an individual more often than would be predicted by random chance
Deletion	A particular kind of mutation: loss of a piece of DNA from a chromosome. Deletion of a gene or part of a gene can lead to a disease or abnormality	Mutation	A permanent structural alteration in DNA. In most cases, DNA changes either have no effect or cause harm, but occasionally a mutation can improve an organism's chance of surviving and passing the beneficial change on to its descendants
DNA	Deoxyribonucleic acid; the molecule that encodes genetic information. DNA is a double-stranded molecule held together by weak bonds between base pairs of nucleotides. The four nucleotides in DNA contain the bases: adenine (A), guanine (G), cytosine (C), and thymine (T). In nature, base pairs form only between A and T and between G and C; thus, the base sequence of each single strand can be deduced from that of its partner	Phenotype	The observable manifestation of a specific genotype
		Polymorphic site	A chromosome site with two or more identifiable allelic DNA sequences. Also called a polymorphic locus
		Polymorphism	Difference in DNA sequence among individuals
		Point mutation	A change in a single base pair
		SNPs	Single nucleotide polymorphism; DNA sequence variation caused by a change in a single nucleotide
Exon	Any segment of a gene that is represented in messenger RNA	Substitution	Replacement of one nucleotide in a DNA sequence by another nucleotide or replacement of one amino acid in a protein by another amino acid
Gene	The fundamental physical and functional unit of heredity. A gene is an ordered sequence of nucleotides located in a particular position on a particular chromosome that encodes a specific functional product (i.e., a protein or RNA molecule)	Transcription	The process by which the genetic information encoded in the gene, represented as a linear sequence of deoxyribonucleotides, is copied into an exactly complementary sequence of ribonucleotides known as messenger RNA
Genome	The full complement of chromosomes, and extrachromosomal DNA coding for cellular proteins, contained within each cell of a given species	Translation	Formation of peptides on ribosomes, directed by messenger RNA

6

Cardiac, Circulatory, and Microvascular Changes in Sepsis and Multiorgan Dysfunction Syndrome

Chris Snowden and Joseph Cosgrove

Most patients who die of sepsis develop a multiorgan dysfunction syndrome (MODS), and outcome from sepsis is strongly related to the number of organs that fail. MODS has varied etiologies, including sepsis, trauma, hemorrhage, burns, myocardial infarction, acute pancreatitis, ischemia-reperfusion injury, and fulminant liver failure. Usually, it follows an overly severe or prolonged systemic inflammatory insult involving activation of components of peripheral blood, complement, and fibrinolytic pathways, leading to the production of a vast array of proinflammatory mediators, such as cytokines, nitric oxide (NO), and endothelins (see Figure 6.1). Concomitant exhaustion of protective, endogenous defence mechanisms (e.g., activated protein C and antithrombin III) is thought to exaggerate the predominance of the proinflammatory environment. This inflammatory state may dissipate within days, but the resultant injury to organ systems can persist, predisposing to organ failure.[1]

In clinical terms, the mainstay of current therapy is centered on ensuring maintenance of global cardiac output and regional organ perfusion in an attempt to prevent secondary organ system injury after the initial inflammatory insult. This approach requires an understanding of the pathophysiology within the cardiovascular system (CVS) during sepsis and the influence of current therapies on such changes.

The vascular endothelium plays an important role at the systemic inflammatory response syndrome (SIRS)-sepsis-MODS interface,[2,3] and it is recognized that dysfunction of the vascular endothelium is an integral step in the initiation of the deleterious microcirculatory changes that create a severe imbalances in oxygen demand, delivery, and use. Subsequent intrinsic cellular derangements predispose to organ dysfunction.

Cardiovascular Effects of Sepsis

Myocardium

Despite a characteristically high cardiac output response to a septic inflammatory challenge, paradoxical myocardial depression is frequently observed (Figure 6.2).[4,5] Clinically, this may manifest as decreased response to fluid resuscitation and catecholamines with biventricular dilation and a reduction in systolic ejection fraction. Myocardial depression is reversible and, in survivors, recovers within 7 to 10 days. The role of diastolic cardiac dysfunction, secondary to reduced left ventricular compliance (particularly relevant in elderly septic patients), is under investigation. Recent studies indicate that myocardial depression may relate to problems with calcium ion (Ca^{2+}) uptake/release from the sarcoplasmic reticulum via voltage-gated Ca^{2+} channels (ryanodine receptors) in the sarcolemma. A reduction of these receptors in the hypodynamic phase of sepsis may reduce the rate of Ca^{2+} release from the sarcoplasmic reticulum, limiting Ca^{2+} availability for interaction with the contractile proteins, thus, depressing systole. Furthermore, there is also a decrease in the rate of uptake from the cytosol back into the sarcoplasmic reticulum, delaying the onset of relaxation and, hence, diastole. The

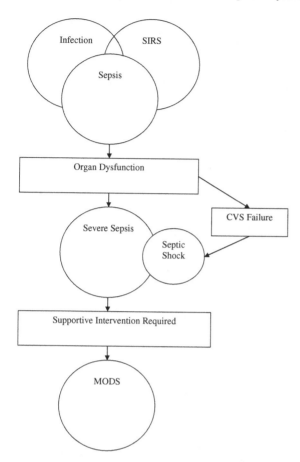

FIGURE 6.1. The infection to MODS cascade.

mechanism underlying the reduction in the number of these Ca^{2+} channels is probably related to systemic myocardial depressant factors, e.g., tumor necrosis factor (TNF)-α and interleukin (IL)-1β, acting either with or independently from NO to interfere with Ca^{2+} release.[4–6]

Systemic Circulation

Sepsis is associated with a widespread systemic vasodilation with reduced reactivity of vasculature.[7] The key instigator of the response is the inflammatory cascade. Because a significant number of potential mediators are released during this response, it has been difficult to determine which specific factors have the greatest vasodilatory influence. Indeed adrenoreceptor activity, alterations in membrane potentials, adenosine triphosphate (ATP)-sensitive potassium (K_{ATP}) channel activity, prostaglandins, thromboxanes, and leukotrienes may all be important. The significant disruption in NO synthesis during sepsis ensures that it remains the most investigated "culprit" for septic-induced vasodilation.[8] However, abnormalities in both vasopressin regulation and the adrenocortical axis may also have a role (Figure 6.3).

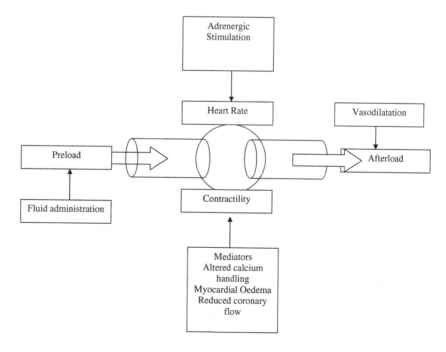

FIGURE 6.2. Influences on myocardial function during a sepsis episode. Note that reduced myocardial contractility is only one effect of sepsis. The overall cardiac response is dependent on circulatory and neurohumoral influences.

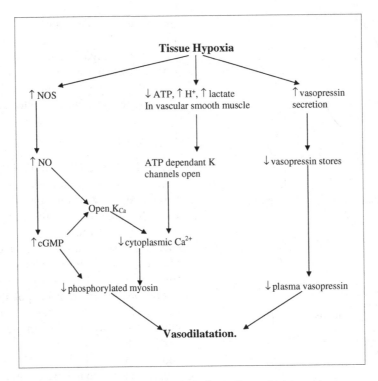

FIGURE 6.3. Proposed mechanisms of vasodilatory shock in sepsis.

NO is produced from L-arginine through the enzyme NO synthase (NOS) Three NOS isoforms exist:

- eNOS: endothelial
- nNOS: neuronal
- iNOS: an inducible form in a number of locations, e.g., macrophages, smooth muscle, and endothelium

eNOS and nNOS are constitutive enzymes often grouped together as constitutive NOS (cNOS). They are concerned with low-output NO pathways involved in homeostatic vascular changes. Under normal physiological conditions, eNOS induces a basal release of endothelial NO, which diffuses to the underlying smooth muscle cells. Guanylate cyclase is activated leading to an elevation of guanosine 3′,5′-cyclic monophosphate (cGMP). This triggers a series of intracellular events culminating in falls in free calcium levels and vascular muscle relaxation. In contrast, several stimuli (associated with inflammation) induce iNOS expression. iNOS stimulation promotes the synthesis of large quantities of NO, its

activation being sustained for several days and being out of the control of negative feedback loops. In experimental sepsis, excessive NO production promotes extensive systemic vasodilation. The evidence for this in human sepsis is less convincing, although clinical trials using NO antagonists (L-arginine analogs) have shown beneficial effects on hemodynamics. However, one large randomized controlled trial (RCT) of an NO antagonist was prematurely stopped because of excess mortality in the treatment group. This suggests that NO has beneficial roles in the response to sepsis; possibly related to the immune response or control of cellular energetics.

Vasopressin acts as a potent endogenous vasoconstrictor at higher concentrations (9–187 pmol/L) than required for its water conservation qualities. In the early stages of shock, plasma vasopressin concentration increases, but, after sustained hypotension there is a rapid reduction in levels. The postulated mechanism is one of depletion of neurohypophyseal stores after profound and sustained baroreceptor stimulation. It is unlikely that vasopressin is the major etiological factor behind systemic circulatory collapse, because initial vasodilation occurs despite increased vasopressin levels. However, the reduction in plasma concentration worsens vasodilatory shock and synthetic analogs of vasopressin are currently being used as therapeutic vasoconstrictors in sepsis, where catecholamine response is reduced.

Regional Circulations

Although global vasodilation is characteristic of sepsis, regional circulations are variously affected.[9] Nonuniformity of blood flow distribution may be detrimental to matching regional oxygen delivery to demand, through "stealing" of blood flow from hypoperfused to well-perfused regions. Reduced perfusion pressure through susceptible tissue beds undoubtedly contributes to both direct and indirect organ injury demonstrated in MODS. Examples of this occur in the hepatomesenteric and renal circulations.

Hepatomesenteric Circulation

The role of hepatomesenteric hypoperfusion in predisposing to MODS has long been debated. Ischemia-reperfusion injury leading to both endothelial and mucosal injury may facilitate endotoxin translocation and mesenteric lymph spread. If endogenous defenses do not remove endotoxin successfully in the mesenteric lymph system, systemic spread may promulgate remote organ injury. In addition, hepatic portal spread can promote both local hepatic injury through complex cellular interactions and worsening of the systemic immuno-inflammatory response. This may provide a secondary inflammatory stimulus for the subsequent progression to MODS—the "second hit theory" (Figure 6.4). The fact that the presence of normal hepatic function before the development of sepsis benefits prognosis supports this hypothesis.[10,11]

Renal System

Alterations in renal artery blood flow, secondary to regional vasodilation or hypovolemia, reduces glomerular filtration rates. Renal tubules (often microscopically normal) have their function impaired through endothelial edema and neutrophil sequestration. Renal tubular dysfunction

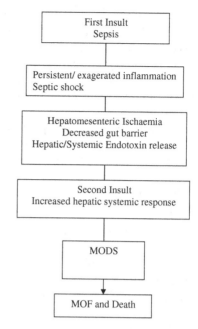

FIGURE 6.4. The "two hit" theory of MODS and multiorgan failure (MOF).

predisposes to systemic fluid, electrolyte, and acid-base imbalances.

Cardiovascular Support in Sepsis

Patients with early sepsis are hypovolemic because of fluid shifts to extravascular spaces. As a result, cardiac output is often initially low. Cardiovascular resuscitation involves early fluid administration and, wherever necessary, vasoactive agents.[12] Early aggressive goal-directed fluid resuscitation and correction of hypotension to normal physiological parameters has a treatment benefit in sepsis and other causes of MODS.[13]

Vasoactive therapies are required when fluid resuscitation has failed to restore adequate organ perfusion. The choice of agent is determined by the relative degrees of impairment of vascular tone and myocardial contractility. Current evidence suggests that norepinephrine (NE) is useful in vasodilatory shock, to increase mean arterial pressure (MAP) through stimulation of α-adrenergic membrane receptors in the peripheral vasculature.[14] Impaired myocardial function may require inotropic support, with dobutamine acting as a β-1 adrenergic agonist. Its β-2 adrenergic effect can reduce MAP, and a vasoconstrictor may also be necessary. Other inotropic agents include dopamine, dopexamine, phosphodiesterase inhibitors, calcium, and digoxin. Newer cardiovascular adjuncts in septic shock include vasopressin and low-dose steroids.

Plasma vasopressin levels fall rapidly in septic shock.[15] Low-dose vasopressin infusions increase MAP, systemic vascular resistance, and urine output in patients responding poorly to catecholamines. There are a number of possible mechanisms of action:

- Plasma concentrations of vasopressin are relatively low. Vasopressin receptors in the vasculature are often available for binding to exogenous hormone.
- Vasoconstrictive effects are increased when autonomic impairment is present secondary to sedation or coma.
- Vasopressin potentiates the vasoconstrictor effect of NE. Plasma NE levels are increased in vasodilatory shock.
- Vasopressin-induced inactivation of K_{ATP} channels in vascular smooth muscle.
- Vasopressin blunts the NO-induced increase in cGMP, which produces vascular dilation.

Vasopressin does, however, have a limited dose response.[16] High rates of infusion (>0.04 U/min) are associated with decreased cardiac output, myocardial ischemia, and renal vasoconstriction.

None of these drugs has been subjected to large RCTs, and the choice of agents, therefore, remains mostly empirical. Therapeutic endpoints are also poorly defined, although the goal-directed study of Rivers et al. does provide some guidance.

Microcirculation and Cellular Considerations in Sepsis

Microcirculation

There is considerable evidence that the primary and major pathophysiological disturbance is sepsis occurs at the microvascular and cellular level. Effective oxygen delivery at tissue level is dependent on adequate regional circulatory flow. In sepsis, oxygen delivery may be reduced by

variations in regional flow distribution or by obstructed capillary flow. Vascular endothelial cells become activated under stressful environments, including sepsis, leading to diminished barrier function, a localized procoagulant state, and expression of adhesion receptors stimulating platelet and leucocyte adhesion.[17] This response is initially protective; adaptive genes synthesize proteins that protect tissues. However, if the stressor persists, the reaction becomes excessive and uncontrolled, with loss of endothelial integrity through direct cellular injury or injury to interconnecting gap junctions. Direct damage to underlying tissues occurs along with capillary obstruction, further exacerbation of tissue hypoxia, and, ultimately, organ dysfunction (see Figure 6.5).[18]

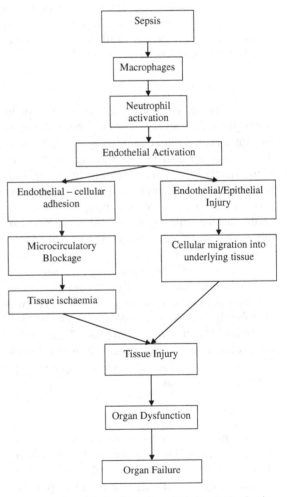

FIGURE 6.5. Integral role of vascular endothelium in the development of organ dysfunction in sepsis.

Cellular Abnormalities

The occurrence of intracellular hypoxemia in sepsis remains controversial. Direct and indirect measurements of intracellular oxygen tensions in sepsis have produced conflicting results and it is likely that both oxygen delivery and consumption failure occur in the septic state. The relative contribution that these deficiencies make to organ failure may vary from tissue to tissue and even within tissues.

Reduced tissue oxygen delivery is the consequence of abnormal control of capillary flow, endothelial damage, and subsequent vascular occlusion. However, in sepsis, the balance between oxygen delivery and cellular demand is also of importance in determining the extent of cellular hypoxia. At a critical level, where demand outstrips supply, an oxygen debt develops and anaerobic metabolism ensues. The progression of tissue hypoxia may also act as a potent inflammatory trigger and a viscous circle of inflammation to tissue hypoxia to inflammation may develop, with the size and duration of oxygen debt contributing to the severity of organ dysfunction. A further consequence of oxygen debt is lactate production. Initially, this is beneficial because it allows high-energy phosphate bonds to be formed during anaerobic metabolism, without the inhibitory affect of accumulation of pyruvate. However, persistent lactic acidosis is a poor prognostic indicator in sepsis. Although tissue hypoxia and anaerobic metabolism are traditionally regarded as the root cause for lactic acidosis, in the context of sepsis, an elevated lactate concentration may also be related to increased lactate production and reduced lactate clearance.

A failure to use oxygen at the cellular level may also occur, leading to tissue dysoxia and "functional" oxygen debt. The role of NO in regulating cell respiration is likely to be of importance.[19–21] Physiological levels of NO probably regulate cell respiration by acting on the mitochondrial cytochrome c oxidase (complex IV) to reduce use of oxygen. Because NO can be competitively displaced by oxygen, the interaction between oxygen and physiological levels of NO protects the cell by reducing oxygen consumption when oxygen levels become low. However, in sepsis, there is inadequate O_2 to displace elevated levels of NO from complex IV. Consequently, the respiratory chain becomes reduced, and the reactive oxygen species, O_2^*, is formed in the mitochondria. The O_2^*, in turn, reacts with free NO to form the peroxynitrite anion, ONOO*. In contrast to the reversible effects of NO on complex IV, ONOO* causes irreversible damage to complexes I and III, lowering the mitochondrial membrane potential and initiating events that cause programmed cell death (apoptosis; see below).

This picture is complicated by a plethora of conflicting studies on mitochondrial function during sepsis. Variability between different tissue defence mechanisms against NO toxicity may also be the cause of contradictory results. However, there is a general consensus that, in the early phase of sepsis, there may be increased mitochondrial function, with a depression of activity later. There is now evidence in humans of mitochondrial dysfunction in a number of tissues during sepsis, including monocytes, intestinal mucosa, liver, and skeletal muscle. The degree of dysfunction seem to be related to the severity of sepsis and, possibly, survival.

Apoptosis and MODS

Apoptosis may be important in both the initiation and recovery from organ failure and sepsis. Apoptosis is an "ordered" form of cell death with specific morphological and biochemical changes, in contrast to necrosis, in which cell membrane disruption results in the release of cellular contents causing inflammation to adjacent tissue. Dying cells lose contact with adjoining cells and disintegrate into membrane-bound fragments that are phagocytosed. In the context of MODS, it seems to be an attempt to remove cells that are a potential threat, including endothelial cells.[22,23] Apoptosis occurs either through removal of "positive" survival signals or the initiation of "negative" cell death signals. IL-2 and neuronal growth factors are examples of "positive" signals. Negative signals can be both internal and external:

- *Internal signals*: The outer membranes of mitochondria express the surface protein bcl-2 that is bound to the protein Apaf-1.

Internal cellular damage leads to the release of Apaf-1. Cytochrome c leaks into the cytoplasm, binding to Apaf-1 and proteases (caspase 9), establishing a chain reaction causing protein digestion, DNA degradation, and phagocytosis.

- *External signals*: These include oxidants, TNF-α and lymphotoxin (TNF-β.) Binding to specific cell membrane receptors activates apoptosis.

Summary

The development of MODS after sepsis is caused by an overriding adaptive host response, integrally related to the prolongation and progression of a proinflammatory insult that overwhelms a counterregulatory protective anti-inflammatory response. Systemic therapy, including catecholamines and vasopressin, are usually required to counteract profound cardiovascular changes, which include both myocardial dysfunction and systemic circulatory vasodilation. In contrast, regional tissue perfusion is difficult to manipulate. Defects in tissue oxygen balance and an inability of oxygen use secondary to dysfunction of mitochondrial enzymatic systems may be an important cause of programmed cell death or apoptosis. Damage to the vascular endothelium is likely to be an important instigator of this response. Until the pathophysiological mechanisms responsible for the progression from severe sepsis to MODS are defined, the treatment of organ failure will remain primarily supportive.

References

1. Singer M, De Santis V, Vitale D, Jeffcoate W. Multiorgan failure is an adaptive, endocrine-mediated, metabolic response to overwhelming systemic inflammation. *Lancet* 2004;364:545–548.
2. Bauer PR. Microvascular responses to sepsis: clinical significance. *Pathophysiology.* 2002;8:141–148.
3. Aird WC. The role of the endothelium in severe sepsis and multiple organ dysfunction syndrome. *Blood* 2003;101:3765–3777.
4. Court O, Kumar A, Parrillo JE, Kumar A. Clinical review: Myocardial depression in sepsis and septic shock. *Crit Care* 2002;6:500–508.
5. Grocott-Mason RM, Shah AM. Cardiac dysfunction in sepsis: new theories and clinical implications. *Intensive Care Med.* 1998;24:286–295.
6. Kumar A, Krieger A, Symeoneides S, Kumar A, Parrillo JE. Myocardial dysfunction in septic shock: Part II. Role of cytokines and nitric oxide. *J Cardiothorac Vasc Anesth.* 2001;15:485–511.
7. Young JD. The heart and circulation in severe sepsis. *Br J Anaesth.* 2004;93:114–120.
8. Vincent JL, Zhang H, Szabo C, Preiser JC. Effects of nitric oxide in septic shock. *Am J Respir Crit Care Med.* 2000;161:1781–1785.
9. Brealey D, Singer M. Multi-organ dysfunction in the critically ill: effects on different organs. *J R Coll Physicians Lond.* 2000;34:428–431.
10. Szabo G, Romics L Jr, Frendl G. Liver in sepsis and systemic inflammatory response syndrome. *Clin Liver Dis.* 2002;6:1045–1066, x.
11. Ring A, Stremmel W. The hepatic microvascular responses to sepsis. *Semin Thromb Hemost.* 2000;26:589–594.
12. Dellinger RP. Cardiovascular management of septic shock. *Crit Care Med.* 2003;31:946–955.
13. Rivers EP, Nguyen HB, Amponsah D. Sepsis: a landscape from the emergency department to the intensive care unit. *Crit Care Med.* 2003;31:968–969.
14. Martin C, Viviand X, Leone M, Thirion X. Effect of norepinephrine on the outcome of septic shock. *Crit Care Med.* 2000;28:2758–2765.
15. Holmes CL, Patel BM, Russell JA, Walley KR. Physiology of vasopressin relevant to management of septic shock. *Chest* 2001;120:989–1002.
16. Holmes CL. Vasopressin in septic shock: does dose matter? *Crit Care Med.* 2004;32:1423–1424.
17. Galley HF, Webster NR. Physiology of the endothelium. *Br J Anaesth.* 2004;93:105–113.
18. Fink MP, Delude RL. Epithelial barrier dysfunction: a unifying theme to explain the pathogenesis of multiple organ dysfunction at the cellular level. *Crit Care Clin* 2005;21:177–196.
19. Brealey D, Karyampudi S, Jacques TS, Novelli M, Stidwill R, Taylor V, et al. Mitochondrial dysfunction in a long-term rodent model of sepsis and organ failure. *Am J Physiol Regul Integr Comp Physiol* 2004;286:R491–R497.
20. Brealey D, Singer M. Mitochondrial dysfunction in sepsis. *Curr Infect Dis Rep.* 2003;5:365–371.
21. Brealey D, Brand M, Hargreaves I, Heales S, Land J, Smolenski R, et al. Association between mito-

chondrial dysfunction and severity and outcome of septic shock. *Lancet* 2002;360:219–223.

22. Hotchkiss RS, Karl IE. Endothelial cell apoptosis in sepsis: a case of habeas corpus? *Crit Care Med.* 2004;32:901–902.

23. Hotchkiss RS, Swanson PE, Freeman BD, Tinsley KW, Cobb JP, Matuschak GM, et al. Apoptotic cell death in patients with sepsis, shock, and multiple organ dysfunction. *Crit Care Med.* 1999;27:1230–1251.

7
Specific Bacterial Infections in the Immunocompetent Patient

Hamad A. Hadi and D. Ashley Price

Standard definitions of sepsis emphasise the similarities in host response to severe infection. However, this "lumping" approach ignores important differences in the clinical presentation and management of certain infectious diseases. This chapter seeks to redress the balance by focusing on a number of important and common bacterial illnesses that may present to the critical care unit.

Meningococcal Disease

Neisseria meningitides, a gram-negative diplococcus, only infects humans. Up to 15% of healthy people carry the organism, but carriage can be as high as 50% in closed groups, such as first-year university students. There are five clinically important serotypes; A, B, C, Y, and W-135. Serotypes A, B, and C account for 90% of cases. Group A epidemics occur in the meningitis belt of Africa. Epidemics with serotype W-135 occurring during the Hajj season can cause major outbreaks, and may be seen in returning pilgrims from Mecca. The quadrivalent vaccine to serotypes A, C, Y, and W-135 is used to prevent disease among travellers to endemic areas, whereas the introduction of meningococcal type C vaccination in the United Kingdom has led to a sustained decline in cases. It is more difficult to produce an immunogenic vaccine against serogroup B; hence, its predominance in the United Kingdom.

The processes of transmission of meningococcal infection, mucosal invasion, and progression to septic shock are not fully understood. Only a fraction of contacts and carriers develop clinical disease. Mucosal barriers and local immunity form initial barriers to bacteremia and these may be compromised by respiratory viral infections. Systemic antibodies against the bacterium are found in approximately 80% of the adult population. Terminal complement deficiency (C5-9), hypogammaglobulinemia, and hyposplenism are risk factors for meningococcal sepsis and, if recurrent infection occurs, then immune defects should be excluded.

Clinical Presentation

N. meningitidis leads to a spectrum of disease, ranging from meningitis and septicemia to septic arthritis and respiratory infections. Chronic infection can also occur. Meningitis accounts for 50% of meningococcal disease, with a mortality of 1 to 5%; 40% have combined meningitis and septicemia and 10% septicemia alone. Meningococcal septicemia is the most severe form of meningococcal disease, with a mortality of up to 40%. The burden of disease occurs in children and young adults. There is strong evidence that outcome in meningococcal disease is related to rapidity of appropriate antibiotic treatment. When meningococcal disease is suspected in the community, antibiotics should, therefore, be administered before transfer to hospital.

Meningitis

The cardinal symptoms of meningitis are fever, headache, photophobia, and neck stiffness. Other systemic symptoms include malaise, nausea, and vomiting, as well as reduced level of consciousness. A purpuric rash typical of meningococcal septicemia is only present in 50% of patients with meningococcal meningitis. Most patients present within 24 to 48 hours of the onset of symptoms and have a good prognosis. Later presentation is associated with increased morbidity and mortality. Partial treatment may mask the clinical features without eliminating the infection. If there is a clinical suspicion of bacterial meningitis, then antibiotics should be started immediately and continued unless an alternative definitive diagnosis is established.

Meningococcal Septicemia

Meningococcal septicemia is characterized by sepsis associated with a purpuric rash (Figure 7.1). Discrete macular erythematous rashes can occur initially, progressing to purpura and then hemorrhagic purpura, but this is not universal. It is essential that the patient is fully examined, because purpura may be scant and confined to areas of pressure. Lesions may occur in the conjunctivae. The patient will often be shocked at presentation, with evidence of multiorgan failure, limb ischemia, and brain hypoxemia. A poor prognosis is suggested by rapid development of purpura, increasing base excess, a high C-reactive protein (CRP) level, and disseminated intravascular coagulation (DIC), with thrombocytopenia and clotting abnormalities. Patients with intractable hypotension may

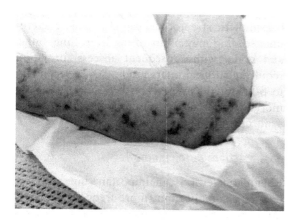

FIGURE 7.1. Purpuric rash in meningococcal septicemia.

have acute adrenal insufficiency secondary to adrenal hemorrhage.

Diagnosis and Management

The primary aim in a patient with suspected meningococcal disease is effective resuscitation and early antibiotic therapy. At initial assessment, cultures of blood, throat, scrapings of purpura, and meningococcal polymerase chain reaction (PCR) should be taken. Approximately 50% of blood cultures and smears from purpuric lesions are positive for meningococci, whereas the sensitivity and specificity of PCR is approximately 90%.

A lumbar puncture (LP) is not required if there is purpura and evidence of septicemia, and an LP should never delay antibiotic administration. Neuroimaging before LP is not necessary in immune-competent patients with normal level of conscious and no focal neurological signs or papilledema. Delaying LP and antibiotic treatment while waiting for brain imaging is associated with increased morbidity and mortality and does not completely prevent the risk of coning. If obtained, cerebrospinal fluid (CSF) examination classically shows low glucose, elevated protein, and a predominant neutrophilia. Gram staining may reveal gram-negative intracellular diplococci. Latex agglutination for meningococcal antigens may be positive. Culture of CSF is often negative if previous antibiotic therapy has been administered.

In suspected meningococcal disease, a third-generation cephalosporin (i.e., Cefotaxime or Ceftriaxone) should be administered as soon as possible. Only if there is a clear history of significant anaphylaxis should chloramphenicol be used as an alternative. In all cases of septicemia, critical care advice should be sought early and there should be a low threshold for intensive therapy unit (ITU) admission, because rapid deterioration often occurs.

Although the role of steroids is defined in cases of pneumococcal meningitis in reducing mortality and neurological sequelae, there is not enough evidence to support its blind use in meningococcal meningitis; however, steroids may help in cases with high CSF protein and white cell counts or visible bacteria.

There is no direct evidence for the efficacy of activated protein C in meningococcal sepsis, but it should be considered in cases of severe sepsis. Multiorgan failure can be treated supportively, whereas limb ischemia may require vasodilator therapy and input from vascular and plastic surgical teams. Limb compartment syndromes can occur and may require escharotomy. Cases of meningococcal disease need to be reported urgently to the local communicable disease control authorities. Prophylactic antibiotics for a patient's close contacts may prevent further cases, but it is not generally necessary for the care team to take prophylaxis.

β-Hemolytic Streptococci

β-Hemolytic Streptococci are gram-positive cocci and important pathogens in humans. They can be grouped with Lancefield antisera into A, B, C, G, or F. Asymptomatic pharyngeal carriage of A, C, and G is common (10–20% of schoolchildren). Group B streptococcus is found in the vagina of up to 30% of women aged 15 to 45 years. Transmission may occur to infants from up to 75% of women colonized with Group B β-hemolytic streptococcus, and may result in neonatal sepsis.

Streptococcus pyogenes or Group A β-hemolytic streptococcus (GAS) is the most important pathogen. It is highly infectious and patients with this infection should be isolated, particularly in the burns and ITU setting.

Pathogenesis

Surface proteins promote colonization. The M protein is particularly important in binding to epithelial cells of the skin and also has antiphagocytic function. Its structure is similar to myosin, leading to induction of cross-reacting antibodies that are probably responsible for rheumatic carditis. A number of exotoxins are produced that can enhance the ability of the organism to spread through tissue. There may also be cardiotoxic (Streptolysin O), nephrotoxic (streptokinase), pyrogenic (Streptococcal pyrogenic toxin), erythrodermic, or act as superantigens causing toxic shock syndrome (TSS).

Clinical Presentations

The clinical manifestations of the Group A, C, and G β-hemolytic streptococci may be grouped into superficial infections, complicated invasive infections, and immunological sequelae (in the case of Group A infection).

Superficial Infections

Pharyngitis

Streptococcal pharyngitis has an incubation period of 1 to 5 days. A sore throat with pharyngeal hyperemia and exudates, a temperature of higher than 38°C, and cervical lymphadenopathy in the absence of cough or viral coryzal symptoms is suggestive of bacterial infection. Peritonsillar abscess can occur and may require surgical drainage. Respiratory embarrassment is very rare and preventable by early initiation of antibiotics and drainage of a peritonsillar abscess. Scarlet fever, which is now rare, presents with a generalized erythematous rash, with sparing of the hands and feet and perioral region. A strawberry tongue also occurs.

Skin Infections

Impetigo produces vesicles and yellow crusting. Cellulitis and erysipelas are deeper infections. Erysipelas results in a well-demarcated area of swollen erythematous skin that is exquisitely painful, often on the face, with constitutional symptoms.

Invasive Infections

β-Hemolytic streptococci can also cause bacteremia, seeding may occur to the liver, lung, bone, and brain, with consequent abscess formation requiring surgical intervention. Sepsis secondary to β-hemolytic streptococci carries a 27 to 38% mortality. Necrotizing fasciitis and toxic shock syndrome are discussed below.

Postinfectious Autoimmune Sequelae

Rheumatic fever is an important but now less common complication (more frequent in developing countries). Presentation is 10 to 25 days after the initial illness. Migratory arthritis occurs in 70% and carditis in 50% of patients. Most of the clinical manifestations are self-limiting apart from the carditis, which can lead to valvular scarring. Glomerulonephritis occurs 1 to 3 weeks after infection with oliguria, hematuria, proteinuria, hypertension, and edema. The prognosis is usually good, but progression to chronic glomerulonephritis and renal failure can occur in 5% of cases.

Diagnosis

Diagnosis of infection is usually by culture from throat swab, pus, and/or blood culture. A throat swab is 95% sensitive in bacterial pharyngitis. A high antistreptolysin titre (ASOT) can be useful, but false positives can occur.

Management

β-Hemolytic streptococci are highly sensitive to penicillin. For pharyngitis, therapy with penicillin V or a macrolide for shorter than 10 days may be followed by recrudescence. Patients with more invasive illness and sepsis initially require high-dose parenteral penicillin (penicillin-allergic patients should be treated with a third-generation cephalosporin), with the addition of clindamycin in those with necrotizing fasciitis, sepsis, or toxic shock syndrome, because it reduces bacterial toxin production. Immunoglobulins have also been shown to be useful in those with invasive infection and sepsis. Rheumatic fever responds to high-dose aspirin or corticosteroids.

Toxic Shock Syndrome

TSS is a toxin-mediated illness that rapidly progresses into shock and multiorgan failure with significant morbidity and mortality. TSS may be mediated by toxins produced by *Staphylococcus aureus* or GAS. TSS is caused by superantigens that directly stimulate up to 20% of CD4$^+$ T cells to produce catastrophic overproduction of Th1 cytokines. Twenty percent of nasal carriers harbor *S. aureus* capable of producing TSS toxin. Staphylococcal TSS can be separated into menstrual, associated with tampon usage, and nonmenstrual, which may be seen postsurgically, postpartum,

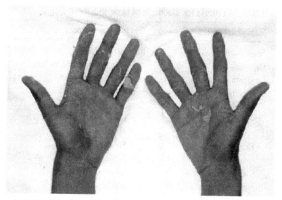

FIGURE 7.2. Desquamation of palms after staphylococcal TSS.

with influenza, and in burns patients. Streptococcal TSS is normally associated with wounds and soft tissue infection. Staphylococcal TSS has a 5% mortality if treated early, but mortality is up to 50% in streptococcal TSS.

Clinical Presentation

The symptoms of TSS consist of fever, headache, malaise, intense myalgia, erythematous macular rash that eventually desquamates (particularly over the palms and soles) (Figure 7.2), mucositis, nausea, vomiting, and diarrhea (a common feature), followed by hypotension and multiorgan failure. DIC with cutaneous hemorrhage may occur.

Diagnosis

Tampon usage and any history of minor trauma should be established.

Diagnosis is suspected clinically and confirmed by fulfilling the consensus case definition (Table 7.1).

Management

Management of TSS is mainly supportive. Thorough examination of the patient for a source of infection is required, with removal of tampons and debridement of any deep-seated infection. Signs of inflammation in surgical wounds infected with TSS strains of *S. aureus* may be slight. Antibiotic therapy should include parenteral

TABLE 7.1. Criteria for diagnosis of toxic shock syndrome

Staphylococcal TSS	Streptococcal TSS
Fever	Isolation of GAS
Hypotension	• From a sterile site (definite)
Desquamating rash	• From nonsterile site (probable)
Organs dysfunction (at least	Hypotension
three)	Organs dysfunction (at least two)
• Renal	• Renal
• Hepatic	• Hepatic
Hematological	• Pulmonary
• GIT	• Hematological
• CNS	• Cutaneous rash
• Musculoskeletal	• Soft tissues necrosis
• Mucous membrane	
No other alternative diagnosis	
with negative blood/CSF	
cultures and negative	
viral/bacterial serology	

ranging between 30 and 70%, hence, its importance.

Clinical Presentation

Most signs of necrotizing fasciitis are nonspecific and delay in diagnosis is common. History of trauma, penetrating injury, or surgical procedure may be elicited, but, in many cases, no precipitating cause is found. Severe pain with systemic symptoms of infection, such as fever and myalgia, occur. Initially, these symptoms are out of proportion to overlying skin changes. With the onset of gangrene and skin breakdown, anesthesia occurs. Examination of the skin may reveal erythematous patches, which commonly progress into purple or grey discoloration and, ultimately, bullous breakdown and gangrene (Figure 7.3). Signs of sepsis and impending shock may only develop later.

flucloxacillin and clindamycin (which may help reduce toxin production). Early use of intravenous immunoglobulins combined with antibiotics may reduce mortality. Activated protein C may be considered according to guidelines.

Necrotizing Fasciitis

Necrotizing fasciitis is a life-threatening deep-seated soft tissue infection that progressively destroys fascia and deeper tissues but initially spares overlying skin. It can occur with gas gangrene or as a primary clinical entity. Necrotizing fasciitis is subdivided into two subtypes based on causative organisms:

1. Type 1 necrotizing fasciitis is polymicrobial and occurs in patients with diabetes, in patients with peripheral vascular disease, and after surgical procedures. Anaerobes, non-A β-hemolytic streptococci, and Enterobacteriaceae (e.g., *Escherichia coli*, *Enterobacter*, and *Klebsiella*) may be involved.

2. Type 2 necrotizing fasciitis is less common and occurs spontaneously or after minor injury. It is caused by GAS, but *S. aureus* may also be isolated. Skin and superficial soft tissue changes are more common in Type 2 necrotizing fasciitis.

Despite its wide publicity, Necrotizing fasciitis remains uncommon, but carries a mortality

Diagnosis and Management

Laboratory markers of infection are present, and creatinine phosphokinase level mirrors progression; however, the diagnosis is ultimately clinical. If early necrotizing fasciitis is suspected clinically, cross-sectional imaging may reveal edema, myonecrosis, and progression of disease along fascia planes. If necrotizing fasciitis is clinically obvious, then imaging is not necessary before proceeding to surgery.

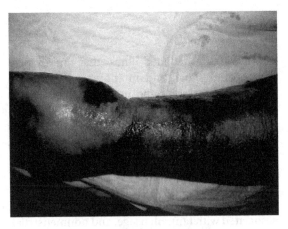

FIGURE 7.3. Late-stage necrotizing fasciitis.

Management of necrotizing fasciitis entails collaboration between infection specialists, intensivists, and surgeons. Surgery is obligatory to debride all infected tissue and prevent or treat compartment syndrome. Repeated surgical intervention and amputation may be necessary. Tissue and fluid cultures are vital to guide antibiotic therapy. The patient should be adequately fluid resuscitated and treated with high-dose broad-spectrum antibiotics. Initial treatment is with a third-generation cephalosporin and clindamycin, but should be tailored depending on culture results.

Surgical Infections

Approximately two-thirds of all nosocomial infections occur in surgical patients. The most common postsurgical infections are at the surgical site (40%), urinary tract (40%), respiratory tract infection (15%), and blood stream (5%). A number of generic factors increase the risk of infection after surgery, including tissue injury, fecal contamination, skin colonization, ventilation, hyperglycemia, age, comorbidity, nutrition, bladder catheterization, and line placement. Events such as immunosuppressive therapy or splenectomy carry specific infection risks.

Antibiotic prophylaxis is common surgical practice and reduces postoperative infections in contaminated wounds. Prophylactic antibiotics should be initiated before procedures and are not usually continued beyond 24 hours after surgery.

Infection control measures are vital to reduce infective complications. Hand washing, preferably with alcohol gel, is the first preventive measure. Minimizing ventilatory support and intensive care stay is essential because these sites harbor resistant organisms. Early removal of catheters, lines, drains, and devices is recommended to reduce the risks of infection. The development of infection with resistant bacteria, such as methicillin-resistant staphylococcus aureus (MRSA) should always be considered in patients in the hospital.

Clinical Presentation

Up to 40% of patients have fever postoperatively, and many of these resolve spontaneously without antibiotics. Noninfectious causes of fever often occur within 3 days of the operation and include deep venous thrombosis (DVT), tissue injury, medication, and atelectasis, but an infectious cause should be sought.

Investigations

Routine investigations should include a full examination, blood cultures, wound inspection and culture, urinalysis, blood gases if hypoxic or septic, and chest X-ray. Further investigation, if the source of fever is not obvious, is governed by the type of operation, but may include ultrasound and/or computed tomographic (CT) scanning looking for abscess formation and exclusion of DVT.

Wound Infection

Wound infections occur after 5 days postoperatively and present with fever, pain, erythema, and sometimes pus discharge. The pathogen involved depends on the procedure. In both clean and contaminated procedures, *S. aureus* is the most common infecting organism (often by endogenous spread). Contaminated procedures are likely to be infected by the normal flora of the viscus entered; consequently, *E. coli*, enterococcus, and anaerobes are common after colonic procedures. In the presence of a purulent discharge, the wound should be explored and pus drained. Necrotizing fasciitis is a rare but serious complication. Flucloxacillin may be used in wound infections after clean procedures. Wound infections after contaminated procedures may require initial empiric treatment with cefuroxime and metronidazole. Particularly if systemically unwell, enterococci should be covered with Amoxicillin. MRSA needs to be considered and treated with antibiotics according to local resistance patterns.

Urinary Tract Infection

Urinary tract infections commonly occur after 3 days, often after catheterization; and present with lower abdominal pain, hematuria, dysuria, and/or frequency. Dipstick and culture of urine is essential. Antibiotic therapy with good gram-negative cover is essential.

Pneumonia

Postoperative pneumonia carries a mortality of approximately 10%. It presents after 3 days postoperatively. Three or more new findings of fever, cough, sputum, chest pain, shortness of breath, and tachycardia indicate postoperative pneumonia. Oxygen therapy, pain relief, and physiotherapy are essential. Pathogens include streptococcus pneumonia, gram-negative pathogens, and anaerobes. Initial antibiotic therapy should be with parental Cefuroxime and metronidazole. Severe illness or lack of response to first-line antibiotics should raise the question of more resistant organisms, such as pseudomonas.

Sepsis

Empiric antibiotic treatment with Cefuroxime and metronidazole may be considered in septic postoperative patients after thorough examination and culture. In the ITU setting, consideration of line infection and further coverage of MRSA with antibiotics such as teicoplanin should be considered.

Suggested Reading

Bernard GR, Vincent J-L, Laterre P-F, et al. The recombinant human activated protein C worldwide evaluation in severe sepsis (PROWESS) study group, efficacy and safety of recombinant human activated protein C for severe sepsis. New England Journal of Medicine 2001;344:699–709.

Cohen J, Powderly WG. Infectious Diseases. 2nd ed. Elsevier, Mosby 2004.

Cunningham JD, Silver L, Rudikoff D. Necrotising Fasciitis: A plea for early diagnosis and treatment. The Mount Sinai Journal of Medicine 2001;68: 253–261.

DiNubile MJ, Lipsky BA. Complicated infections of skin and skin structures: when the infection is more than skin deep. Journal of Antimicrobial and Chemotherapy 2004;53 suppl S2:ii37–ii50.

Keh D, Sprung CL. Use of corticosteroid therapy in patients with sepsis and septic shock: an evidence-based review. *Critical Care Medicine* 2004;32(11 Suppl):S527–533.

Mandell GL, Bennett JE, Dolin R. Principles and practice of Infectious Diseases. 6th ed. Elsevier, Churchill Livingstone 2005.

McCormick JK, Yarwood JM, Schlievert PM. Toxic shock syndrome and bacterial superantigens: An update. Annual Reviews of Microbiology 2001; 55,77–104.

Norrby-Teglund A, Ihendyane N, Darenberg J. Intravenous immunoglobulin adjunctive therapy in sepsis, with special emphasis on severe invasive group A streptococcal infections. Scandinavian Journal of Infectious Diseases 2003;35 (9):683–689.

Van de Beek D, de Gans J, McIntyre P, Prasad K. Steroids in adults with acute bacterial meningitis a systemic review. Lancet Infectious Diseases 2004; 4:139–143.

8
Infection in the Immunocompromised Patient

Michael H. Snow and Nikhil Premchand

Infection in the immunocompromised not infrequently leads to intensive therapy unit (ITU) admission and, for example, up to 40% of bone marrow transplant patients require intensive care unit admission after transplantation. The outcome for the immunocompromised patient requiring critical care is notoriously poor, with less than a 2% hospital survival quoted in some older case series. However, recent advances in treatment (e.g., early use of noninvasive ventilation) coupled with resuscitation that is more aggressive has led to improvements in outcome. In addition the longer-term survival in some conditions that were once uniformly fatal (e.g., advanced HIV disease) has been transformed in recent years.

Increased Susceptibility to Infection

Increased susceptibility to infection may occur because of defects in skin or mucosal integrity, impaired nonspecific immune defenses (phagocytes, cytokines, and complement cascade), or failure of specific cell-mediated or humoral immune function. In addition, intravenous devices, intravenous feeding, gastric acid suppression, antibiotic use, and metabolic problems, such as diabetes or the effects of alcohol, may further increase the risk of infection and influence the range of microorganisms involved. Splenectomy caused by the impaired removal of opsonized bacteria and reduced antibody production against polysaccharide antigens predisposes to overwhelming sepsis with capsulated organisms, particularly streptococcus pneumoniae.

Specific predisposing factors are each associated with a particular spectrum of organisms with limited overlap, and this may be of importance in considering the microbiological differential diagnosis and, hence, the "best guess" antibiotic therapy (see Table 8.1).

Granulocytopenia

Granulocytopenia is caused most commonly by the treatment of hematological and solid organ malignancies, but can also result from inherited conditions, such as chronic granulomatous disease, which causes abnormal neutrophil phagocytosis. It has been long recognized that both the degree and duration of specific predisposing factors determine risk of infection. The risk of infection is only slightly increased until the granulocyte count drops below 500 cells/μL and then rises rapidly as the count falls toward zero (see Figure 8.1). Qualitative defects of phagocyte function also predispose to infection. Infections are often initially localized (oropharynx, perineum, lung, gastrointestinal tract, etc.) (see Figure 8.2), but bacteremia/septicemia is common. Local epidemiology and resistance patterns influence first-line antibiotic regimens.

Antibody Deficiency

Antibody deficiency occurs in congenital immunodeficiency and secondary to myeloma, chronic lymphatic leukemia, and immunosuppressive/cytotoxic therapy. It predisposes to bacterial infection, particularly with capsulated organisms.

TABLE 8.1. The range of infections associated with different specific immune defects and additional factors that may influence these

Nature of immunocompromise	Common organisms	Reasons
Granulocytopenia	Gram-negative organisms • Escherichia coli • Pseudomonas aeruginosa • Klebsiella spp.	Endogenous flora
	Gram positive organisms: • Staphylococcus aureus • Coagulase-negative staphylococci (CNS) • Enterococci • Streptococcus viridans group	Skin flora Wider use of central venous access catheters Use of prophylactic quinolone antibiotics More intensive chemotherapy regimes resulting in bacterial translocation from the gut
	Invasive fungal infections • Candida spp. • Aspergillus spp.	Use of broad spectrum antibiotics, parenteral nutrition, and steroids
Humoral immunodeficiency	Streptococcus pneumoniae Haemophilus influenzae Neisseria meningitidis Mycoplasma spp.	
Splenectomy	Streptococcus pneumoniae Haemophilus influenzae Neisseria meningitides Salmonella spp. (especially Sickle cell disease)	Encapsulated polysaccharide antigens

Respiratory infections and bacteremias are common.

Complement Deficiency

Complement deficiency involving the later components (C5 to C9), particularly predispose to neisserial infections, and may lead to recurrent meningococcal disease.

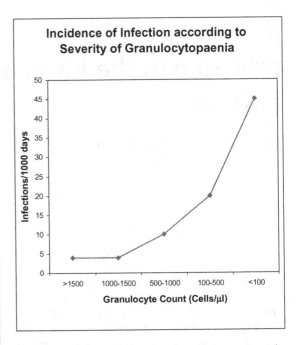

FIGURE 8.1. The effects of severity of granulocytopenia on the incidence of infections in acute leukemia patients during induction chemotherapy. (Adapted from Bodey et al., 1966.)

Defects of Cell-Mediated Immune Function

Defects of cell-mediated immune function predispose to the reactivation of latent infections, such as toxoplasma, herpes group viruses, Pneumocystis jiroveci (formally carinii) and opportunist infections, such as mycobacterium avium complex. Many of these organisms are

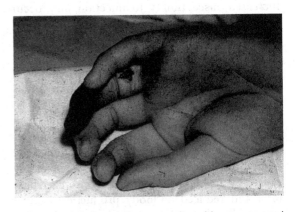

FIGURE 8.2. Soft tissue infection complicated by gangrene and pseudomonas septicemia secondary to carbimazole-induced granulocytopenia.

intracellular pathogens. Impaired T-cell function also occurs with AIDS, lymphoma, transplant immunosuppression, cytotoxics, renal failure, and inherited immunodeficiencies.

Immediate Management

The initial investigation of potentially infective syndromes in the compromised patient must be guided by the nature of the specific predisposing factors, the organ system involved, and guidance from investigations.

In patients with conditions associated with virulent bacterial infections and septicemia (e.g., granulocytopenia or splenectomy) after initial cultures (blood ×3, urine, swabs of throat, skin lesions, sputum culture, wound swabs, etc.), broad-spectrum antibiotic therapy must be initiated immediately. Most bacteremias will be confirmed by three blood cultures incubated for 72 hours. Intravenous access sites and devices must be carefully assessed and cultures taken. However, the significance of positive blood cultures taken from intravenous lines can be difficult to interpret. Polymerase chain reaction (PCR) for cytomegalovirus (CMV), Epstein-Barr virus (EBV), and other viruses should be considered.

Imaging (ultrasound scan, computed tomography [CT], magnetic resonance imaging [MRI], chest x-ray) should be preformed early and there should be careful clinical examination of the oropharynx, perineum/perianal areas, fundi, and pelvis. Central lines may need replacing and their tips should be cultured for bacteria. If there are focal abnormalities, then targeted sampling should be performed, e.g., bronchoalveolar lavage for Pneumocystis jiroveci (formerly carinii) pneumonia (PCP) in an HIV-positive patient with cough and interstitial shadowing on chest x-ray.

If immunocompromised patients with nonspecific features of infection and no organ dysfunction fail to respond to antibacterial therapy, the possibility of fungal disease becomes more likely, and a therapeutic trial of antifungal therapy, particularly directed at aspergillus, must be considered.

To illustrate some of these principles, we will discuss granulocytopenic sepsis, infection in humeral immunodeficiency, CMV infection in solid organ transplantation, and PCP in HIV disease.

Management of Granulocytopenic Sepsis

Granulocytopenic sepsis is a medical emergency and may present with nonspecific fever, a sepsis syndrome, or focal organ dysfunction. Recognized early and treated promptly with antimicrobial and supportive measures, mortality from neutropenic sepsis can be significantly reduced.

Up to 80% of bacterial infections arise from the patient's endogenous flora (see Table 8.1). The commonest bacteria isolated are gram-negative bacteria, such as *Escherichia coli*, *Pseudomonas aeruginosa*, and *Klebsiella sp.*, but during the last 20 years, there has been an increase in gram-positive infections.

Fever is common in granulocytopenic patients, and it has been estimated that 80% of patients with neutropenia for longer than 1 week will have a febrile episode, and 60% of these febrile episodes are thought to be infectious in origin. Other causes for fever in granulocytopenic cancer patients are the underlying malignancy, prescribed medication, the use of blood products, and inflammatory conditions, such as hematomas or thromboembolic disease.

After cancer chemotherapy, fever can occur at any time, however, the median time to fever is 9 to 10 days from initiation of chemotherapy or approximately 3 days from the development of granulocytopenia. Fever may be the only symptom, and a thorough clinical examination is vital to look for other clues regarding the source. Despite thorough investigation, a causative organism may only be found in a minority of cases.

A patient's risk of developing neutropenic sepsis can be stratified as low, intermediate, or high, and treatment modified according to the risk assessment:

- Low-risk patients are those likely to have short periods of neutropenia (3–5 d) and with no significant comorbidities or barriers to adherence, and have rapid access to medical care in the event of deterioration. These patients may be treated as outpatients with oral antibiotics.

- Intermediate-risk patients are those undergoing stem cell transplantation for solid tumors or lymphoproliferative malignancies when the granulocytopenic period may be prolonged (8–13 d). In the event of fever, they should receive intravenous antibiotics initially.
- High-risk patients are those who are predicted to have neutropenia for greater than 14 days or who have significant comorbidities or hemodynamic instability. These patients have significantly higher risk of mortality, and require intravenous antibiotics and very careful monitoring.

Management can be divided into the following phases:

- Preventative: prophylactic and preemptive
- Empirical treatment of suspected infection
- Treatment of established infection

It is important that local antibiotic guidelines are followed and that staff liaise closely with the microbiology or infectious diseases teams to target the use of antimicrobial therapy according to local antibiotic resistance patterns.

Preventative measures must be taken to reduce the risk of infection. From a nursing point of view, single rooms for patients and diligent hand washing for healthcare workers and visitors are recommended. There is no good evidence that wearing a gown, gloves, or a face mask have any effect on reducing infection rates. High-efficiency particulate air (HEPA) filtration is useful to protect high-risk patients from invasive pulmonary aspergillosis. This is especially important if there is demolition or building work, because this releases large numbers of aspergillus conidia into the environment.

Fluoroquinolone prophylaxis for patients predicted to be at high risk of granulocytopenia results in a reduction in the risk of infection with gram-negative bacteria and *Staphylococcus aureus*. Although there is no reduction in the incidence of febrile episodes, infection-related mortality is reduced. Prophylactic antifungal agents in patients at risk of severe granulocytopenia, with agents such as fluconazole, reduce subsequent invasive fungal infection. Preemptive therapy used when an individual has been found to be colonized with bacteria or fungi may prevent

invasive fungal infection, but has a limited evidence base.

Empirical antimicrobial therapy must be used at the onset of symptoms while awaiting microbiological results. There are multiple regimens available (see Table 8.2), but several factors need to be taken into account in decision making (see Table 8.3). If a causative organism is subsequently identified, then drug therapy should be adjusted, if necessary.

Patients with severe and prolonged neutropenia with negative cultures, who remain febrile after 5 to 7 days of empirical antibiotics, are at high risk (up to 20%) of invasive fungal infections. Empirical antifungal treatment reduces the risk by 50 to 80%, and reduces mortality from fungal

TABLE 8.2. Empirical antibiotic therapy for febrile neutropenic episodes

Regimen type	Antimicrobial type	Examples
Monotherapy	Antipseudomonal penicillin + β-lactamase inhibitor	Piperacillin/tazobactam Ticarcillin/clavulanate
	Carbapenem	Imipenem/cilastin Meropenem
	Fluoroquinolone	Ciprofloxacin Levofloxacin
	3rd or 4th generation Cephalosporin	Ceftazidime Ceftriaxone Cefixime
Combination therapy	Antipseudomonal β-lactam	Piperacillin Carbapenem Cephalosporin (with antipseudomonal activity)
	PLUS Aminoglycoside	Gentamicin Tobramycin Amikacin
	OR Fluoroquinolone	Ciprofloxacin Levofloxacin
Empirical antifungal agents to be added to above regimes if necessary:		
	Class	Examples
	Polyene (fungicidal)	Amphotericin B (and lipid formulations thereof)
	Azole (fungistatic)	Itraconazole Fluconazole Voriconazole
	Echinocandins (fungistatic)	Caspofungin

Source: Adapted from Rosser and Bow, 2004.

TABLE 8.3. Factors to consider in choosing an empirical antibiotic regime in granulocytopenic patients with fever

Patient renal function and allergies
Known or suspected infecting organism
Predominant organism on the unit involved
Resistance patterns
Activity and spectrum of antimicrobial activity
Route of administration
Drug pharmacokinetics and interactions
(Cost)

infections by 23 to 45%. A number of different antifungal agents can be used (see Table 8.3).

The median time to defervescence in low-risk patients is 2 to 3 days and 4 to 6 days in high-risk patients. Empirical therapy should be continued for at least 7 days after defervescence in the first 5 days of empirical therapy. Patients who remain febrile should continue antibiotics for 5 days after the neutrophil count is above 0.5×10^9 cells/L or if the neutrophil count does not recover for a minimum of 14 days.

Adjunctive therapy such as granulocyte colony stimulating factor (G-CSF) can also be used to boost the neutrophil count. Although there is no evidence of decreased mortality, hospital stay and duration of antibiotic therapy are reduced. Although costly, usage of hemopoietic growth factors should be considered for high-risk patients.

Humoral Immunodeficiency

Humoral immunodeficiencies can be caused by inherited B-cell defects or acquired as a result of chronic lymphocytic leukemia (suppression of B-cell function), multiple myeloma (reduced B-cell numbers) or the nephrotic syndrome (urinary immunoglobulin loss).

The commonest inherited immunodeficiency is **IgA deficiency**, which affects as many as 1 in 300 people. The phenotypic presentation is variable, with the majority remaining asymptomatic, but some patients developing atopy, recurrent sinopulmonary infections, or autoimmune disorders.

Children with **X-linked hypogammaglobulinemia** usually remain well until the age of 6 to 9

months because they are protected by passive immunity from the mother. Later they often develop recurrent infections with *Streptococcus pneumoniae*, *Haemophilus influenzae*, meningococci, and *Mycoplasma sp.*, resulting in sinusitis, otitis media, and pneumonias. By the time the true diagnosis is made, irreversible bronchiectasis may already have developed.

Common variable hypogammaglobulinemia manifests later in life, generally between the ages of 15 and 25 years. Patients often have normal numbers of B cells with very low levels of secreted immunoglobulin, probably as a result of impaired T helper cells. This can result in similar infections to patients with agammaglobulinemia and should be considered in all patients presenting with bronchiectasis.

Investigation and Management

If a humoral defect is suspected as a result of recurrent infections, initial investigations include total Ig levels, IgG subclasses and functional antibody levels. Immunoglobulin replacement therapy, usually intravenously, should be administered every 3 to 4 weeks to prevent the development of encapsulated bacterial infections and the late sequelae of these. Live vaccinations should be avoided in patients with agammaglobulinemia or hypogammaglobulinemia and inactivated vaccines are unlikely to be effective.

CMV Infection in Transplant Recipients

There are numerous different infections that can affect recipients of solid organ transplants and these vary depending on the organ transplanted, time since transplant, and degree of immunosuppression (see Figure 8.3). The degree of immunosuppression required to prevent rejection depends on the organ transplanted (bone marrow transplant requires more immunosuppression than heart or lung, which require more than liver, which requires more than renal) and how well matched the donor and recipient are. Immunosuppression is generally most intense in the early months after transplantation. The immunosuppressive

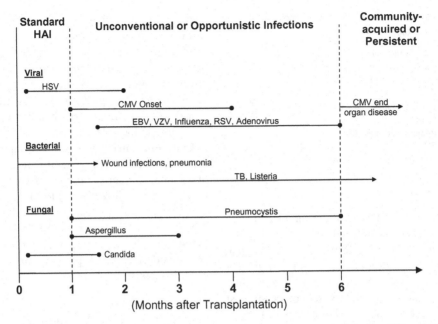

FIGURE 8.3. The relationship between the spectrum of opportunist infections and the time since transplantation in solid organ transplant recipients. HAI, hospital-acquired infection; HSV, herpes simplex virus; EBV, Epstein-Barr virus; VZV, varicella-zoster virus; RSV, respiratory syncytial virus; TB, tuberculosis. (Adapted from Fishman and Rubin, 1998.)

agents belonging to the calcineurin inhibitor classes (cyclosporine A and tacrolimus) have predominant T-cell effects; corticosteroids (prednisolone) and the nucleotide synthesis inhibitors (azathioprine and mycophenolate mofetil) cause both T- and B-cell suppression.

Posttransplantation Infections

Infections in the posttransplantation period can be divided into three phases—in the first month, 1 to 6 months, and longer than 6 months after transplantation. In the first month, 95% of infections are the same as in patients undergoing comparable surgery without immunosuppression, but it is possible for the recipient to become infected (especially with viruses) by the transplanted organ. Immunosuppression can result in the recrudescence of infection present in the recipient before surgery; this is why it is critical to eradicate all infections before transplantation, if possible.

From 1 to 6 months after transplantation, viral infections are very common, with CMV causing approximately two-thirds of febrile episodes in this period (see below). The incidence of PCP is 15% in this period and, therefore, appropriate antibiotic prophylaxis, usually with co-trimoxazole to prevent opportunistic infection, is very important.

Six months after transplantation, most cases of chronic viral infections and opportunistic infections occur in patients with graft rejection and, hence, more intense immunosuppression.

CMV Infection

CMV is a human herpes virus and is one of the most important pathogens in transplant recipients. Problems can either arise through primary infection from the graft, infection from blood products, or, rarely, from contact with an individual with active infection. Alternatively, immunosuppression and inflammatory processes resulting in tumor necrosis factor (TNF)-α production can result in reactivation of latent virus in the recipient. Hence, knowing the serostatus of donor and recipient can help predict the likelihood of clinically significant infection. If both donor and recipient are CMV seronegative, then the use of CMV-negative blood products can significantly reduce the risk of infection. If the recipient is CMV seropositive, then the risk of reactivation when antiviral prophylaxis is not used is 80% in the setting of bone marrow transplantation. For CMV-negative recipients receiving organs from seropositive donors in the same setting, the risk of infection is 40%.

In the immunocompetent host, primary infection is usually asymptomatic but in the immunocompromised, CMV infection has a wide variety of clinical effects. These range from a mild illness with lymphadenopathy, fever, and malaise (and possible lymphocytosis, leucopenia, and thrombocytopenia) to end organ dysfunction (pneumonitis, encephalitis, hepatitis, or enterocolitis). Interstitial pneumonitis affects 10 to 15% of bone marrow transplant patients and, untreated, has a mortality of 80%. Clinical symptoms are of dyspnea and fever with associated hypoxia and chest X-ray changes. After solid organ transplant, the main effects are seen in the transplanted organ; lung transplant recipients can develop pneumonitis; liver transplant recipients, hepatitis; and renal transplant recipients, CMV-associated renal artery stenosis. CMV infection itself has an immunosuppressive effect and can lead to superinfection with other organisms, such as invasive aspergillus, gram-negative bacteria, and Pneumocystis.

Diagnosis

Diagnosis of CMV has been revolutionized with the advent of molecular techniques, such as PCR. There is still a role for serological tests to determine previous exposure and, therefore, stratify risk, but PCR is the mainstay for surveillance (the onset of viremia precedes the development of end organ disease) and the investigation of active disease. PCR can also be used on biopsies to detect end organ disease.

Treatment

Several antiviral drugs with activity against CMV are available. The most efficacious drugs are ganciclovir (which can only be administered intravenously) and valganciclovir, an ester of ganciclovir that is absorbed orally. Should resistance develop to ganciclovir, then foscarnet or cidofovir can be used, but both these agents are nephrotoxic.

Two strategies can be adopted for the treatment of CMV. Patients can be administered prophylaxis with ganciclovir during high-risk periods, but there is the potential for nephrotoxicity and myelotoxicity. The alternative is preemptive therapy started with the development of PCR positivity, administering 5 mg/kg ganciclovir twice daily for 14 days (or shorter if two consecutive PCR tests become negative). If end organ disease is present, then at least 14 days of intravenous therapy should be used followed by maintenance therapy with valganciclovir. If cidofovir is used, adequate prehydration and probenecid should be used to minimize nephrotoxicity.

Pneumocystis Jiroveci Pneumonia

Pneumocystis jiroveci (previously carinii) was considered a protozoan, but genetic analysis shows it to be a yeast. It is a classic opportunist organism, rarely, if ever, causing infection (as opposed to colonization) in the immunocompetent patient. Early in the HIV epidemic, PCP was the AIDS indicator disease in two of three patients. It rarely occurs with CD4 counts of greater than 200. In patients of known HIV status, under follow-up, prophylaxis with co-trimoxazole (480 mg daily or 960 mg three times a week), 50 mg dapsone daily, or 500 mg azithromycin daily before the CD4 count falls to 200 will prevent almost all infections. However, patients with undetermined HIV status continue to present with advanced immunodeficiency and PCP. Patients who lack recognized risk factors for HIV acquisition may suffer considerable delays in diagnosis and may, therefore, develop severe respiratory insufficiency. When a patient presents in respiratory failure it is very important that full respiratory and ITU support are given because, if the patient survives the acute illness, the longer-term prognosis on antiretroviral therapy is very good.

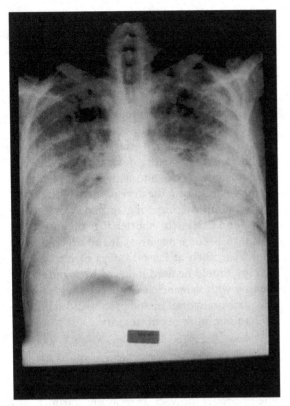

FIGURE 8.4. Chest X-ray in a patient with severe PCP showing predominantly lower zone interstitial change. The patient had severe respiratory failure.

Presentation

The presentation of PCP is subacute, with symptoms developing over several days to weeks. Features include night sweats, dry cough often induced by temperature change or respiratory irritants, and malaise. Dyspnea on exertion and then at rest supervenes. Systemic symptoms and weight loss are common.

Important clinical clues include potential HIV exposure, other symptoms of immunodeficiency (e.g., previous shingles or thrush), or signs such as oral hairy leucoplakia. Often, despite evidence of impaired gas transfer (cyanosis, and tachypnea) chest examination findings are unremarkable. Unless sexual, drug, and travel histories are part of standard clinical assessment, missed diagnoses and inappropriate treatment will occur.

Diagnosis

In patients of known HIV status, respiratory symptoms are likely to be reported earlier. They should be screened by resting pulse oximetry and, if normal, exercise oximetry. A fall of oxygen saturation by 5% or more or to below 90% indicates the need for further investigation. In early PCP, plain chest radiology may be normal, even when gas transfer is impaired. High-resolution CT scanning, however, is very sensitive and will demonstrate a ground-glass interstitial pattern. At later stages, chest x-ray typically shows bibasal perihilar shadowing with an interstitial pattern (see Figure 8.4). Pneumothorax may occur and pneumatoceles may be seen. Effusions and hilar lymphadenopathy occur but are uncommon. In patients who have had nebulized pentamidine as prophylaxis, infection may be localized to the upper lobes.

If PCP is suspected, early initiation of therapy is indicated. Diagnostic respiratory samples (bronchioalveolar lavage [BAL] or induced sputum) will remain positive for at least 48 hours after specific antibiotic therapy is begun. Fluorescein-linked monoclonal antibodies against Pneumocystis (see Figure 8.5) enable the diagnosis to be confirmed on spontaneously expectorated sputum occasionally, on induced sputum in approximately 80% of cases, and in bronchoalveolar lavage samples in greater than 90% of cases. Transbronchial biopsy is now only occasionally required and may be complicated by hemorrhage or pneumothorax.

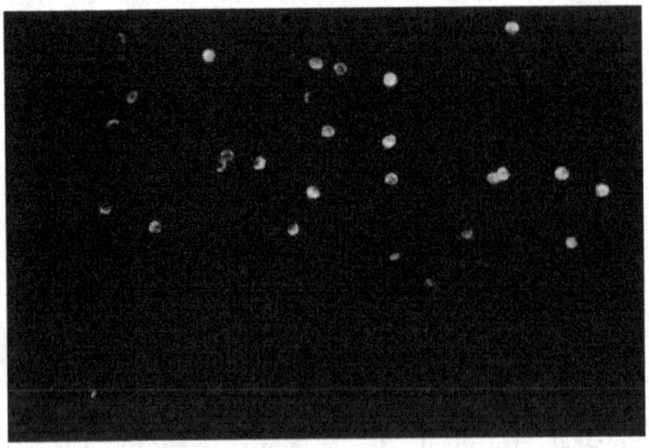

FIGURE 8.5. Immunofluorescent monoclonal antibody staining of BAL specimen in a patient with PCP.

Respiratory Support

In patients with hypoxia and respiratory distress, it is important to recognize that bronchoscopy and BAL, and, less frequently, sputum induction, may precipitate respiratory insufficiency requiring respiratory support with continuous positive airway pressure (CPAP), noninvasive intermittent ventilation (NIV), or intubation and ventilation.

Treatment

The standard therapy for PCP is high dose co-trimoxazole (120 mg/kg in four divided doses) initially parenterally except in mild disease. Bone marrow toxicity may be reduced by concomitant folinic acid. Treatment should be continued for a total of 3 weeks and then secondary prophylaxis initiated as for primary prevention. For patients who are sulfonamide allergic or not responding to first-line treatment, alternate regimens include clindamycin and primaquine, dapsone and trimethoprim, atovaquone, or intravenous pentamidine.

In patients with significant hypoxia, steroid therapy (with prednisolone or methyl prednisolone up to 80 mg twice daily, tailing down over 10 d) reduces the risk of respiratory failure and shortens the severely symptomatic stage at only slight risk of precipitating other problems such as thrush or herpes simplex. It is not uncommon for patients to have some deterioration in lung function during the first few days of treatment. Pneumothorax needs to be watched for and treated with aspiration or chest drainage as appropriate. During high-dose co-trimoxazole therapy, full blood count and biochemistry needs to be monitored regularly, and toxicity or hypersensitivity may necessitate treatment changes.

Antiretroviral therapy should be deferred until completion of the therapeutic co-trimoxazole course. Secondary prophylaxis should be continued until a substantial period of immune reconstitution (CD4 greater than 200) and viral suppression (viral load less than 50) has been achieved. Most patients (80%) who develop hypersensitivity reactions to co-trimoxazole can be successfully desensitized using an escalating dose of the drug.

Suggested Reading

Bodey GP, Buckley M, Sathe YS, et al. Quantitative relationships between circulating leukocytes and infection in patients with acute leukaemia. *Ann Intern Med* 1966;64(2):328–340.

Calandra T, Holland SM. Infections in the Immunocompromised Host. In: Cohen J, Powderly WG, et al. eds. Infectious Diseases. 2nd ed. Edinburgh: Mosby; 2004; pp 97–114.

Fishman JA, Rubin RH. Infection in organ transplant recipients. *N Engl J Med* 1998;338:1741–1751.

Gagnon S, et al. Corticosteroids as adjunctive therapy for severe Pneumocystis carinii pneumonia in the acquired immunodeficiency syndrome. A double-blind, placebo-controlled trial. *N Engl J Med* 1990; 323:1444–1450.

Mandell GL, Bennett JE, Dolin R, eds. Principles and Practice of Infectious Diseases. 6th ed. Philadelphia: Elsevier; 2005;268:306–312.

Rosser SJ, Bow EJ. Infections in neutropenic hosts. In: Loeb M, Smieja M, Smaill F, eds. Evidence-based Infectious Diseases. London: BMJ Publishing Group; 2004; pp 241–256.

9
Severe Infections in the Returning Traveler

Jacob P. Wembri and Matthias L. Schmid

The ease of air travel has made spread of infectious agents a global problem. There are a multitude of tropical diseases ranging from benign viral illnesses to highly contagious and life-threatening diseases and it is important to take a detailed clinical and travel history from the ill returning traveller, especially in the first 4 weeks of return.

The following article will concentrate on three major disease areas:

1. Falciparum malaria, which remains a very important infection because delays in diagnosis may be fatal.
2. Emerging or reemerging infections, such as severe acute respiratory syndrome (SARS) or avian influenza, have a serious impact on patients but also on staff, especially in a critical care environment.
3. There are highly contagious infections, such as viral hemorrhagic fever (VHF), with high mortality, which present occasionally in the United Kingdom. It is important to be familiar with those diseases and their common presentations to be able to deliver the best care in an environment safe for both patients and staff. Recognizing these diseases and instigating appropriate management protocols are essential for patient management and protection of staff.

Severe and Complicated Malaria

There are four human malarial parasites; *Plasmodium falciparum, p. ovale, p. vivax*, and *p. malariae*. Severe and complicated malaria is caused by *p. falciparum* and will be dealt with here. Malaria is estimated to cause 200 million clinical cases and approximately 1 million deaths mainly in sub-Saharan Africa each year. More than 75% of deaths are in children younger than 5 years old. In adults, severe falciparum malaria is more commonly seen in populations vulnerable because of war or famine, during pregnancy, and in nonimmune travellers to malarial endemic areas. In the United Kingdom, severe malaria is most often seen in travellers who have not taken chemoprophylaxis, those who present late, and those with underlying medical conditions. Malarial chemoprophylaxis does not provide absolute protection and, therefore, malaria must be excluded as a cause of febrile illness in all travellers returning from endemic areas.

Clinical and Laboratory Features

The incubation period for falciparum malaria is at least 7 days. Patients commonly present within 2 weeks of return to the United Kingdom. Occasionally, however, they may present later; up to 3 months after return, especially if chemoprophylaxis has been taken. The clinical presentation of severe malaria is very variable depending on the immunity of the infected person. The most common presentation is with fever, chills, malaise, headache, myalgia, anorexia, nausea, diarrhea, dry cough, impaired consciousness, and convulsion (cerebral malaria). They may be pale, dehydrated, jaundiced, hypotensive, confused, and may have tender hepatosplenomegaly. Some of the false localizing features (e.g., significant diarrhea) can cause diagnostic confusion. Table 9.1

TABLE 9.1. Modified WHO criteria for severe malaria in the nonimmune adult

Criteria for severe and complicated malaria in adults
Decreased level of consciousness, Glasgow Coma Scale (GCS) <11 of 15
Generalized convulsion
Extreme weakness
Shock (algid malaria), blood pressure <90/60 mmHg
Hypoglycemia, blood glucose <2.2 mmol/L
Acidosis, blood bicarbonate <15 mmol/L, pH <7.3
Renal impairment, blood urea >17 mmol/L
Anemia, hemoglobin <8 g/L, hematocrit <20%
Hyperbilirubinemia (>43 μmol/L)
Hemoglobinuria (black-water fever)
Disseminated intravascular coagulation
ARDS/pulmonary edema (>32 breaths per minute)
Hyperparasitemia (>2% asexual P. falciparum in nonimmune patients)
Late presentation (>4 d of fever)
No malaria prophylaxis

lists the criteria for severe and complicated malaria. In nonimmune patients, severe malaria can occur without heavy parasitemia.

The most common laboratory abnormality is thrombocytopenia. Other abnormalities include anemia, deranged clotting, hyponatremia with or without renal impairment, raised liver transaminases, raised bilirubin, metabolic acidosis, and low blood glucose and hypoalbuminemia.

Remember: patients who present with fever lasting longer than 4 days and/or who have not taken any malaria prophylaxis may be severely ill, requiring critical care input at an early stage.

Pathogenesis

Several mechanisms are involved in the pathogenesis of complicated malaria.

Parasitized red blood cells (pRBC) develop "knobs" on the cell membrane, which adhere to capillary endothelial cells. In addition, the red cells may conglomerate around pRBC, causing a phenomenon called "rosetting." The end result is sequestration of pRBC, causing blood flow obstruction and tissue anoxia. Release of cytokines, such as tumor necrosis factor (TNF), also have a role in the pathogenesis of cerebral malaria. Anemia is caused by bone marrow suppression and red cell destruction, either directly or immune mediated. Acidosis is a marker of severity of disease caused by lactic acid release from hypoperfusion and cytokine release.

Management

Management should include specialist advice from Regional Infections Disease Units. Severe falciparum malaria (Table 9.1) should be managed in the high dependency unit (HDU) or intensive therapy unit (ITU). The definitive diagnosis is made by finding *P. falciparum* on a blood film (Figure 9.1). A rapid antigen test specific for *P. falciparum* is available, which may help in the initial management of the patient while waiting for the microscopy result. An assessment of severity must be performed (see Table 9.1). Recommended treatment regimens are shown in Table 9.2. Artemisinin-based combination therapy has now been shown to be as effective as quinine in treating severe and complicated malaria, and intravenous artesunate has been shown to be more efficacious with reduced complications and mortality. In the United Kingdom, quinine is still recommended by the British Infection Society guidelines, but intravenous artesunate may be used in certain circumstances, although it is not easily available. In some parts of the world, such as Southeast Asia, there is widespread quinine resistance and so these new drugs are of great importance.

Exchange transfusion can be used in heavy parasitemia (>10%) to increase parasite clearance, although there is no randomized trial for this form of therapy. Introduction of intravenous artesunate may reduce the need for plasma exchange transfusion because it reduces parasitemia much

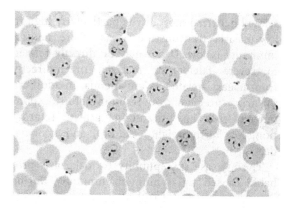

FIGURE 9.1. High parasitemia in a case of severe falciparum malaria.

TABLE 9.2. Management of severe falciparum malaria

Intravenous quinine in 500 mL 5% dextrose over 4 hours; loading dose (only if not taken quinine or mefloquine), 20 mg/kg; maintenance dose, 10 mg/kg every 8 hours (consider intravenous artesunate if available)

Oral quinine (when patient stable and able to swallow); 600 mg every 8 hours for a total of 7 days

200 mg doxycycline daily or 450 mg clindamycin every 8 hours (in pregnancy) to complete 7 days coadministered with quinine

Quinine-resistant malaria: intramuscular artemether, loading dose of 3.2 mg/kg bolus and 1.6 mg/kg daily maintenance; or intravenous artesunate, loading dose of 2.4 mg/kg and 1.2 mg/kg daily maintenance, changing to oral formulation when the patients is stable and able to complete 7 days plus doxycycline or clindamycin

General management, electrocardiogram monitoring during intravenous treatment, oxygen therapy, careful fluid balance, blood sugar monitoring, observations every 4 hours

more rapidly than quinine. Patients may develop acute renal failure and severe lactic acidosis, requiring hemofiltration or hemodialysis. Fluid management is crucial because dehydration is common but overly liberal infusion can cause pulmonary or cerebral edema. Empiric antibiotics may be necessary when sepsis cannot be excluded. Occasionally, patients may develop hypoxic events or cardiomyopathy after severe malaria and these should be considered when recovery, despite parasite clearance, does not proceed as expected.

Severe Malaria in Pregnancy

Pregnancy is a risk factor for severe malaria in endemic areas as well as for visitors. Malaria-related maternal mortality is particularly high in low-malaria transmission areas where natural immunity is low. Death may be directly related to severe falciparum malaria, or indirectly from malaria-related severe anemia. In addition, malaria in pregnancy can result in stillbirth, spontaneous abortion, intrauterine growth retardation, low birth weight (<2.5 kg), and neonatal death.

The increase susceptibility to malaria is probable caused by waning of immunity during pregnancy compared with nonpregnant women. In areas of low transmission and epidemics, pregnant women are prone to malaria infection

developing severe disease because of lack of protective immunity. Additionally nonimmune pregnant travellers are at the highest risk. pRBC are sequestrated in the placenta, resulting in impairment of function. Management of severe malaria in pregnancy requires multidisciplinary input from obstetricians, pediatric intensivists, adult intensivists, and infectious diseases specialists, but should follow otherwise the same management as outlined above (Table 9.2). Particular problems during treatment in pregnancy are hypoglycemia from both the infection and quinine, and pulmonary edema. Quinine is recommended in severe malaria with clindamycin (doxycycline is contraindicated). Artemisinin derivatives are currently not recommended because the teratogenic risk in humans has not been fully assessed. Other drugs used commonly for symptom control, such as nonsteroidal anti-inflammatory drugs (NSAIDs), are relatively contraindicated in pregnancy.

Remember: falciparum malaria in pregnancy is worse in the nonimmune patient and may be associated with serious disease to both mother and child and require early critical care input.

SARS as an Example of an Emerging Contagious Disease with High Morbidity and Mortality

SARS is caused by SARS-associated coronavirus (SARS-CoV), a novel coronavirus that first caused human illness in the Southern Chinese province of Gaungdong toward the end of November 2002. It may have originated as a zoonosis. It spread to Hong Kong in February 2003, when an infected doctor fell ill while staying at a hotel. From this index case, the disease spread in the local population as well as to other Asian countries, Canada, the United States, and Germany. The mortality rate ranged from 0% in mild cases to 50% in those with severe disease. The World Health Organization (WHO) announced an end of SARS transmission in July 2003, by which time, there were more than 8000 probable cases of SARS and 774 deaths in 29 countries, with a case fatality rate of 10%. A significant number of medical and paramedical staff were affected, particularly those involved in airways management. Introduction of new rigorous infection control mechanisms in acute and critical care setting were necessary. Since July 2003, there have been no new outbreaks of infections, but this new disease entity has the potential to reemerge and exemplifies the problems faced by clinicians in the coming years.

Clinical and Laboratory Features

The incubation period of SARS ranges from 2 to 16 days (median, 10 d). The clinical features range from mild respiratory symptoms to very severe disease with pneumonia and respiratory failure. SARS commonly presents with fever, chills, rigors, myalgia, sore throat, and headache followed by a nonproductive cough and breathlessness. In the second week of illness, watery diarrhea may occur and chest signs worsen. Signs may be minimal initially and progress to consolidation and pleural effusion. Approximately 20% of patients develop acute respiratory distress syndrome (ARDS), necessitating ventilatory support, and a high proportion will develop nosocomial sepsis and multiorgan failure with a high mortality rate.

The clinical case definition of suspected, probable, and confirmed cases in the event of reemergences of SARS is listed in Table 9.3. The diagnosis of SARS is made by either isolation of virus by cell culture or by reverse transcriptase (RT) polymerase chain reaction (PCR), or by four-fold rise in antibody titers.

Laboratory findings at presentation include anemia, lymphopenia, thrombocytopenia, hyponatremia, hypocalcemia, raised lactate dehydrogenase (LDH) levels, and elevated liver transaminases. The raised LDH level indicates severity.

An abnormal chest X-ray is present in up to 80% of patients at onset of fever, initially as unilateral patchy shadowing, which progress to bilateral involvement. Early computed tomographic (CT) scans show subpleural focal consolidation.

Management

It is vital to consider SARS as a possible diagnosis during an active outbreak here or elsewhere in the world. Symptomatic "probable" cases of

TABLE 9.3. UK SARS case definition, if SARS reemerges (once verified by WHO)

Suspect case of SARS

Respiratory illness requiring hospitalization characterized by:
• Fever (>38°C) and
• Cough or breathing difficulty and
• One or more of the following exposure during the 10 days before onset of symptoms:
  ◦ Close contact with a suspect or probable case of SARS
  ◦ History of travel to area with recent local transmission of SARS
  ◦ Residing in an area with recent local transmission of SARS
  ◦ History of exposure to laboratories that have retained SARS virus isolates and/or diagnostic specimens from SARS patients

Probable case of SARS
A suspected case with: • Radiological evidence of infiltrates consistent with pneumonia or respiratory distress syndrome (RDS) or • Autopsy finding consistent with the pathology of pneumonia or RDS without an identifiable cause • No alternative diagnosis to fully explain their illness
Confirmed case of SARS

Symptoms and signs that are clinically suggestive of SARS and:
• Laboratory evidence of SARS-CoV infection based on one or more of the following:
  ◦ PCR positive for SARS-CoV (validated method) from at least two different specimens (respiratory, stool)
  ◦ Same clinical specimen collected on two or more occasions
  ◦ Two different assays or repeat PCR using new RNA extract
Seroconversion by enzyme-linked immunosorbent assay (ELISA) or immunofluorescence assay (IFA)
• Positive antibody test during convalescent phase after negative test during acute illness
• Four-fold rise in antibody titers between acute and convalescent sera Virus isolation
• Isolation in cell culture of SARS-CoV from any specimen, plus
• PCR confirmation using validated method

Source: Lim et al., 2004.

SARS require urgent assessment and negative pressure isolation facilities. Clear guidance for admission to a critical care unit needs to be established within each hospital setting. Empiric antibiotic should be used to cover community-acquired pneumonia because clinical presentation is nonspecific and rapid laboratory diagnosis is difficult.

During the SARS outbreak in 2003, ribavirin, a broad-spectrum antiviral was used widely but there is no randomized control trial to assess its efficacy. However, in the absence of alternative agents, this should be considered early in management. Systemic steroids are suggested on the basis of experience in Hong Kong and Toronto as part of a three-pronged approach (Table 9.4). The rationale was that, despite decreases in SARS-CoV viral load and rises in SARS-specific IgG toward the third week of illness, clinical deterioration was observed, probably related to cytokine release induced by the virus. Currently interferon has no use in clinical practice.

Up to 30% of SARS patients required intensive care admission. Of these, 10 to 20% needed mechanical ventilation. Noninvasive ventilation was used widely and was estimated to prevent intubation and mechanical ventilation in two-thirds of severe cases, but infection risk associated with aerosol generation is greatest in noninvasive intermittent ventilation (NIV) and during intubation.

Control of Infection

Healthcare workers (HCWs) are at great risk of contracting SARS, and it was estimated to have affected up to 30% of staff in the Hong Kong outbreak, with occasional fatalities. In particular, HCWs involved in airways management, such as critical care staff, are at high risk of exposure. After the SARS outbreaks in 2003, each hospital

TABLE 9.4. Modified treatment protocol for probable or confirmed SARS

Standard treatment protocol for SARS in adult patients (Hong Kong)
Antibiotic (according to local protocol for community-acquired pneumonia) • 500 mg levofloxacin daily, intravenously or orally; or • 500 mg clarithromycin twice daily plus • 375 mg Amoxiclav (amoxicillin and clavulanic acid) three times daily, orally Antiviral (10–14 d) • 400 mg Ribavirin every 8 hours intravenously for 3 days, or when stable; then • 1200 mg Ribavirin twice daily Corticosteroid regime (21 d) • 1 mg/kg methylprednisolone every 8 hours for 5 days; then • 1 mg/kg methylprednisolone every 12 hours for 5 days; then • 0.5 mg/kg prednisolone twice daily for 5 days; then • 0.5 mg/kg prednisolone daily orally for 3 days; then • 0.25 mg/kg prednisolone daily orally for 3 days; then • Stop • Pulsed 500 mg methylprednisolone twice daily intravenously for 2 days (if worsening) Ventilation

TABLE 9.5. Infection control in hospital management of SARS

Patient to wear surgical mask continuously unless on face mask for oxygen
Admit to a negative-pressure room (single room if unavailable)
Transfer to designated center with facility for isolation (negative pressure)
Ensure HCWs adhere to control measures; gown, gloves, goggles or visors, respirator masks (EN149:2001, FFP3), and strict hand washing
Inform hospital infection control, regional disease control center (CCDC), designated SARS infectious disease unit, and maintain list of all staff in contact with the patient
All staff to be vigilant for symptoms after contact with patient and not to turn up for work if develop symptoms up to 10 days after exposure
Visitors to be restricted (except for immediate family)

should have developed their own protocols to prevent disease transmission. It is, however, important to realize that infection control measures should be applied to every patient. United Kingdom recommendations are shown in Table 9.5.

Other Issues Relating to New or Emerging Infections

The critical care staff are under increased stress during the management of a highly infectious patient and require close monitoring and support. SARS reminds us that there are other potentially serious infections that require all members of staff to be on high alert.

Viral Hemorrhagic Fever

VHF encompasses a wide range of viruses that cause febrile hemorrhagic illness with high case fatality rates. There are four main viruses, which are predominantly zoonotic but can cause human disease (Table 9.6). They are highly infectious agents that can spread by direct contact with blood, secretions, organs, or other bodily fluids of infected persons.

Pathogenesis

Each VHF may present a different disease spectrum, but there are common pathophysiological processes. There is vascular endothelial damage

TABLE 9.6. Main features of VHF

Disease/virus	Family Genus	Vector/route of transmission	Geographical area	Mortality rate
Ebola hemorrhagic fever	*Filoviridae* *Filovirus*	Unknown Human to human Nosocomial spread	Sub-Saharan Africa	25–90%
Marburg hemorrhagic fever	*Filoviridae* *Filovirus*	Unknown Human to human Nosocomial spread	Sub-Saharan Africa	25–90%
Lassa fever	*Arenaviridae* *Arenavirus*	Rodent excreta Human to human	Western Africa	2–15%
Crimea-Congo hemorrhagic fever	*Bunyaviridae* *Nairovirus*	Tick bite Blood of infected animal Human to human	Eastern Europe, Asia, Africa	15–30%

caused by both direct viral injury and indirect effects or from inflammatory cytokines and immune activation. In addition, there is disruption of coagulation system, leading to disseminated intravascular coagulation.

Clinical and Laboratory Features

VHF is considered in a patient with fever who has visited a known endemic area or has been in contact with a suspected or confirmed case up to 21 days before onset of symptoms. Table 9.7 summarizes the risk category for VHF. Clinical presentation of VHF ranges from mild nonspecific symptoms to severe life-threatening manifestations. In severe disease, capillary leakage leads to shock. Bleeding may be extensive from gums, puncture sites, and orifices. Multiorgan failure occurs commonly, including encephalopathy, renal, and hepatic failure.

Laboratory specimens are extremely biohazardous and need to be sent in secure containers to specific Category 4 laboratories. Common initial abnormalities include raised hematocrit, leucopenia, thrombocytopenia, raised transaminases, and disseminated intravascular coagulation. Diagnosis is made by isolation of the virus or raising antibody levels.

It is important that treatable infections that are more common, such as malaria and typhoid, are considered and treated or excluded.

Management

Because VHFs carry high mortality and can spread by close contact, it is paramount that suspected cases are managed in a center with the appropriate expertise and isolation facilities. In the United Kingdom, there are only two such units that use Trexler isolators for patient containment (Royal Free Hospital, London; and Newcastle General Hospital, Newcastle upon Tyne; Figure 9.2). It is, therefore, important to consider the diagnosis of VHF in patients carefully and seek advise early to prevent spread of infection. Supportive treatment and management of complications is important in all cases of VHF. Ribavirin is beneficial in the treatment of Lassa fever and Crimean-Congo hemorrhagic fever.

TABLE 9.7. Risk categorization for VHF

Minimum (febrile patient)
- Not been in endemic area before onset of illness; or
- Onset of illness longer than 21 days after being in endemic area or contact with suspected or confirm case of VHF

Moderate (febrile patient)
- Been in endemic area during the 21 days before onset of illness (no additional risk factors); or
- Been to adjacent area during the 21 days before onset of illness, and have severe disease, e.g., multiorgan failure, with no alternative cause

Severe (febrile patient who has been to an endemic area during the 21 d before illness)
- Stayed in house longer than 4 hours with confirmed or strongly suspected case of VHF
- Nursed or cared for confirmed or strongly suspected case of VHF
- Laboratory or HCW likely to come in contact with body fluid, tissue, or dead body of VHF patient or animal
- Febrile patient not been to endemic area but in contact with confirmed or strongly suspected case or secretions/clinical specimens from such a case

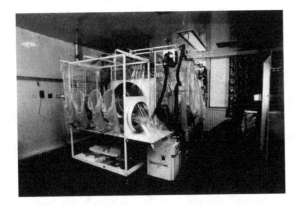

FIGURE 9.2. Trexler Unit for management of patients with VHF. (Newcastle General Hospital.)

Summary

Modern travel broadens the spectrum of diseases imported to the United Kingdom. Tropical or subtropical diseases may be brought back well within their incubation period, adding to diagnostic difficulties. Failure to take an adequate travel history is the commonest cause of "missed" diagnoses. Falciparum malaria, SARS, and VHF are examples of infections with high morbidity and mortality rates, often requiring critical care support. These diseases are uncommon and the general intensivist requires regular updates of knowledge, and procedures for infection control are essential. Up-to-date management will lead to better outcomes for patients and will also reduce the risks to staff. In the future, we may see more patients with tropical illness because of rising travel to increasingly exotic destinations and more travel by at-risk groups, such as the elderly and those with preexisting medical problems.

Suggested Reading

British Infection Society. Algorithm for the initial assessment and management of malaria in adult (draft). 2004; *www.britishinfectionsociety.org.*

Crowcroft N, Brown D, Gopal R, Morgan D. Current management of patients with viral haemorrhagic fevers in the United Kingdom. *Eurosurveillance* 2002;7(3):44–48.

Greenwood BM, Bojang K, Whitty CJM, Targett GAT. Malaria. *Lancet* 2005;365:1487–1498.

Health Protection Agency. Interim guidelines for action in the event of a deliberate release: Viral Haemorrhagic fevers. *HPA-Colindale* 2003, Version 2.1. www.hpa.org.uk/infections/topics_az/deliberate_release.

Ksiazek TG, Erdman D, Goldsmith CS, et al. A novel coronavirus associated with severe acute respiratory syndrome. *N Eng J Med* 2003;348: 1953–1966.

Lim WS, Anderson SR, Read RC. Hospital management of adults with severe acute respiratory (SARS) if SARS re-emerges-updated 10 February 2004. *J Infect* 2004;49:1–7.

Mahanty S, Bray M. Pathogenesis of filoviral haemorrhagic fevers. *Lancet Infect Dis* 2004;4:487–498.

McIntosh HM, Olliaro P. Artemisinin derivatives for treating severe malaria. The Cochrane database of systemic reviews 2000;2:CD000527.

Peiris JSM, Chu CM, Cheng VCC, et al. Clinical progression and viral load in a community outbreak of coronavirus-associated SARS pneumonia: a prospective study. *Lancet* 2003;361:1767–1772.

Richmond JK, Baglole DJ. Lassa fever: epidemiology, clinical features, and social consequences. *Br Med J* 2003;327:1271–1275.

So LKY, Lau ACW, Yam LYC, et al. Development of a standard treatment protocol for severe acute respiratory syndrome. *Lancet* 2003;361: 1615–1616.

Solomon T. Viral haemorrhagic fevers. In: Cook GC, Zumla A, eds. Manson's Tropical Diseases. London, Saunders 21st ed, 2003.

South East Asia Quinine Artesunate Malaria Trial (SEAQUAMAT) group. Artesunate versus quinine for treatment of severe falciparum malaria: a randomised trial. *Lancet* 2005;366:717–725.

World Health Organisation. Severe and complicated malaria. *Transaction of the Royal Society of Tropical Medicine and Hygiene* 2000;94(Suppl 1):S1–S90.

10
Antibiotic Prescribing Including Antibiotic Resistance

Debbie Wearmouth and Steven J. Pedler

The broad selection of antimicrobials now available is vital in treating serious infections in the critically ill. However, it also raises the potential for misuse of these agents, and increasing microbial resistance is occurring. It is, therefore, crucial that antimicrobial use is rationalized by informed and careful prescribing.

Important Antibiotics and their Use in the Critical Care Setting

There are numerous antibiotics available for use in the United Kingdom, although most hospitals have a limited formulary, stocking a small number of antibiotics from each major group. To mention every available antibiotic is beyond the scope of this article and, therefore, only those most commonly used in critical care will be discussed. Where possible, the antibiotic group name is used, with examples given or important differences noted. Table 10.1 shows the classification of antibiotics important in intensive care and their sites of action.

Penicillins

Amoxicillin has activity against gram-positive bacteria, such as streptococci and enterococci, and against many gram-negative bacteria, notably *Haemophilus influenzae, Escherichia coli, Salmonella, Shigella,* and *Proteus mirabilis.* However, they are inactivated by penicillinases produced by some of these organisms and, therefore, their empirical use in septic patients is limited. Flucloxacillin is penicillinase-stable and has antistaphylococcal activity as well as activity against streptococci, but lacks activity against gram-negative organisms. Antipseudomonal penicillins used in combination with a β-lactamase inhibitor (e.g., piperacillin-tazobactam or ticarcillin-clavulanate) have a broad spectrum of activity including many gram-positive, gram-negative, and anaerobic bacteria. They are therefore useful in the intensive care unit (ITU) for the empirical treatment of lower respiratory tract, urinary tract, intra-abdominal, and skin infections, and septicemia.

Cephalosporins

Cephalosporins are broad-spectrum antibiotics used for the treatment of septicemia, pneumonia, meningitis, intra-abdominal, and urinary tract infections. The second-generation cephalosporin, cefuroxime, has activity similar to amoxicillin but is not susceptible to inactivation by staphylococcal β-lactamase and, therefore, also has antistaphylococcal activity (but not against methicillin-resistant *Staphylococcus aureus* [MRSA]) as well as a range of gram-negative bacilli. Third-generation cephalosporins have broader gram-negative cover than the second-generation agents, but lose some antistaphylococcal activity. Cefotaxime is recommended for the treatment of meningococcal meningitis and ceftazidime has antipseudomonal activity.

Carbapenems

Carbapenems are broad-spectrum agents that are useful in the empirical treatment of sepsis and

TABLE 10.1. Classification of antibiotics commonly used in critical care units

Antibiotic group	Site of action	Subgroup	Important examples
β-lactams	Inhibition of cell wall synthesis and remodelling	Penicillins	Benzylpenicillin, penicillin-V
			Amoxicillin
			Flucloxacillin
			Piperacillin, ticarcillin
		Cephalosporins	Cefuroxime
			Cefotaxime, Ceftazidime
			Cefpirome
		Carbapenems	Meropenem, imipenem, ertapenem
		Monobactams	Aztreonam
		β-lactamase inhibitors[a]	Clavulanic acid, tazobactam, sulbactam[a]
Glycopeptides	Inhibition of cell wall cross-linking		Vancomycin, Teicoplanin
Aminoglycosides	Inhibition of protein synthesis		Gentamicin, amikacin, tobramycin
Macrolides	Inhibition of protein synthesis		Erythromycin, clarithromycin
Lincosamides	Inhibition of protein synthesis		Clindamycin
Oxazolidinones	Inhibition of protein synthesis		Linezolid
Quinolones	Inhibition of DNA supercoiling		Ciprofloxacin,
			Moxifloxacin
Nitroimidazoles	Mechanism uncertain		Metronidazole

[a]These are β-lactam agents but possess inadequate activity in their own right and, therefore, must be combined with another agent, such as amoxicillin or piperacillin. They act by inhibiting some of the β-lactamase enzymes produced by bacteria and rendering the organism susceptible again to the agents they are combined with.

other serious infections, and for the treatment of gram-negative sepsis caused by multiresistant organisms. Meropenem is also licensed for the treatment of central nervous system infections. Ertapenem is licensed for treating abdominal and gynecological infections and for community-acquired pneumonia, but, unlike the other carbapenems, is not active against *Pseudomonas* or *Acinetobacter* species.

Aminoglycosides

Gentamicin, amikacin, and tobramycin all have activity against some gram-positive and many gram-negative bacteria, including *Pseudomonas*. They must be administered parenterally and are excreted via the kidney, therefore, can accumulate in impaired renal function. Indications include septicemia, endocarditis, acute pyelonephritis, biliary-tract sepsis, hospital-acquired pneumonia, and neutropenic sepsis. Amikacin is more stable than gentamicin to enzymic inactivation and is, therefore, reserved for treatment of infections caused by gentamicin-resistant organisms. Tobramycin is most commonly administered nebulized

for pseudomonal respiratory infections, particularly in patients with cystic fibrosis.

Fluoroquinolones

Several agents are available, but the most frequently used in critical care is ciprofloxacin, which is moderately active against gram-positive organisms but has a broad gram-negative spectrum, including *Pseudomonas* species. It is indicated for respiratory tract infections (except pneumococcal pneumonia), urinary tract infections, gastrointestinal infections, and septicemia caused by sensitive organisms.

Glycopeptides

The agents vancomycin and teicoplanin are active against gram-positive aerobic and anaerobic bacteria. Because of their potential for renal toxicity, they are reserved for serious gram-positive infections and in the treatment of endocarditis, particularly when caused by MRSA. They are most commonly used in the ITU setting for invasive device-related infections, and sepsis caused by MRSA. Vancomycin can also be administered orally for pseudo-membranous colitis caused by *Clostridium difficile*.

Linezolid

Linezolid is active against aerobic and some anaerobic gram-positive organisms, including MRSA and glycopeptide-resistant enterococci (GRE). However, because of serious potential side effects of hematopoietic disorders, notably thrombocytopenia, its use should be limited to infections caused by these resistant organisms or when alternative agents are not tolerated. It is licensed for the treatment of pneumonia and complicated skin and soft tissue infections.

Principles of Choice of Antimicrobial Agent

Once a diagnosis of infection has been made, there are a number of factors to consider before an antibiotic is chosen. These include both the likely pathogen and its antibiotic susceptibility, and a number of host factors.

Factors Relating to the Pathogen

In many cases, a focal infection is apparent and antibiotics can be chosen to target pathogens that are most likely to occur at that site. For a condition such as community-acquired pneumonia, a small number of organisms cause the vast majority of cases—*S. pneumoniae, H. influenza, Moraxella catarrhalis*, and the so-called atypical pathogens, including *Mycoplasma pneumoniae, Chlamydia pneumoniae*, and *Legionella* species. This permits prediction of likely antibiotic susceptibilities and, in this case, the development of national guidelines for treatment.[1] In other patients, any previous positive cultures may be helpful and, in the absence of this, knowledge of the organisms prevalent in the local unit can be taken into consideration.

In sepsis of unknown origin, once relevant cultures have been taken, broad-spectrum empirical antibiotics are generally required while further investigations are underway. Detailed history and examination are key to identifying the most likely source of infection and guiding empirical antibiotic choice. Risk stratification to identify patients at high risk of colonization with resistant organisms should be undertaken.[2] These risk factors include prolonged length of stay in hospital, previous antibiotic courses, and the presence of invasive devices, for example, central venous cannulae, endotracheal tubes, and urinary catheters.[2]

The antibiotic must not only have an appropriate spectrum for the likely pathogens but also must reach the site of infection at concentrations sufficient to inhibit growth. This is particularly important when considering protected sites such as the central nervous system, when the requirement to cross the blood-brain barrier limits the choice of appropriate agents.

If a pathogen is identified; once relevant susceptibility testing has been performed, it may be possible to rationalize initial treatment to a more targeted regimen. This strategy is thought to help reduce the risk of developing resistance,[2] as well as reducing potential side effects for the individual.

Host Factors

The patient must be able to tolerate the chosen agent(s), which must be administered at an appropriate dose for that condition. A history of previous adverse drug reactions is important, although differentiation must be made between allergic reactions, which would prohibit further use of the drug, and side effects, which may not. General considerations, such as the age of the patient, may change the choice of antibiotic; for example, in elderly patients, it may be desirable to avoid cephalosporins, which increase the risk of *Clostridium difficile* infection. In young women, it is important to determine if they are pregnant because many antibiotics would be contraindicated. Renal and hepatic function may not only influence the choice of antibiotic but also the dose that is required.

Principles for Prescribing Antimicrobial Agents

It is important that a sufficient dose for the site and type of infection is administered. For example, meropenem may be administered at a higher dose

TABLE 10.2. Recommended *minimum* duration of treatment in immunocompetent patients in critical care settings

Condition	Minimum recommended duration
Bacteremia with removable treatable focus (e.g., abscess, contaminated intravenous line)	7 days
Pneumococcal pneumonia	7 days or until afebrile for more than 3 days
Atypical pneumonia	10 days
Meningococcal and *Haemophilus* meningitis	7 days
Pneumococcal meningitis	10–14 days
Listeria / Group B streptococcal / gram-negative meningitis	14 days
Pyelonephritis	10 days
Pelvic inflammatory disease	14 days
Pseudomembranous colitis (*C. difficile*)	10 days

in central nervous system infections compared with other infections. Underdosing can lead to therapeutic failure and also to development of resistance.[2]

In the ITU, most antibiotics are administered by the intravenous route, but the patient may be able to take the medication orally or via a nasogastric tube. For example, antibiotics such as ciprofloxacin and linezolid have excellent oral bioavailability.

The duration of treatment not only depends on the type of infection and infecting organism but also the clinical response of the patient. Certain infections have a minimum recommended treatment course; see Table 10.2 for details. Markers of infection, including the peripheral white cell count and C-reactive protein, are helpful in monitoring response.

Certain antibiotics have a narrow therapeutic range and require drug-level monitoring to ensure that levels are not toxic. Aminoglycosides, including gentamicin, tobramycin, and amikacin, should have blood levels and renal function monitored throughout treatment. Regimens vary from hospital to hospital, and local protocols should be adhered to. Vancomycin also requires monitoring of levels, and recent draft guidelines for the treatment of MRSA infection advocate checking teicoplanin levels to ensure therapeutic doses are being achieved.[3]

Antibiotics are often prescribed empirically on the ITU, usually because urgent treatment is required before the results of culture and/or susceptibility testing are known. Table 10.3 gives some suggested empirical regimens for common conditions in critical care, but it must be emphasized that the choice of antibiotics for any unit depends on knowledge of local drug formularies and especially the common microbial flora encountered in that unit.

TABLE 10.3. Suggested empirical antimicrobial regimens for common conditions

Infection	First-line therapy	Alternative
Acute bacterial meningitis	Cefotaxime	Chloramphenicol[a] (in penicillin-hypersensitive patients)
Urinary tract infection or pyelonephritis	Cefuroxime[b]	Ciprofloxacin
Intra-abdominal sepsis or biliary sepsis	Cefuroxime[b] plus metronidazole[c]	Piperacillin-tazobactam OR meropenem
Pneumonia—community acquired	Cefuroxime[b] plus clarithromycin	Moxifloxacin
Pneumonia—hospital acquired	Piperacillin-tazobactam	Meropenem
Necrotizing fasciitis	Amoxicillin plus ciprofloxacin plus metronidazole[d]	Meropenem[d]
Neutropenic sepsis	Piperacillin-tazobactam plus aminoglycoside	Meropenem
Sepsis of unknown origin	Cefuroxime[b] plus metronidazole	Consult a microbiologist or infectious diseases physician

[a]Meropenem is another alternative in patients not allergic to penicillin, particularly in cases after penetrating injury or neurosurgery.
[b]Cefotaxime or ceftriaxone are acceptable alternatives to cefuroxime.
[c]Amoxicillin should be added in severe sepsis or in proven infection with enterococci, which are always resistant to the cephalosporins. Also, consider adding an antifungal such as fluconazole in patients not responding to antibacterial treatment.
[d]The selection of appropriate agents depends on the predisposing factors in the patient. For example, the suggested first-line regimen is particularly suitable for necrotizing fasciitis developing after abdominal surgery. In patients with suspected or proven Group A streptococcal infection, a regimen containing clindamycin would be potentially advantageous.

Antibiotic Resistance

Definition

There is no consensus regarding how to define antimicrobial resistance. However, an infecting organism is considered susceptible when it is inhibited by a concentration of the drug that is achievable at the site of infection.[4] For practical reasons, resistance is defined in vitro by the use of "breakpoint" concentrations, usually defined as the lowest concentration of the antibiotic that inhibits 90% of bacterial growth under defined conditions (the minimum inhibitory concentration 90 [MIC90]). However, a laboratory result does not always translate directly into clinical response. From a clinical standpoint, it must be understood that a susceptible or resistant result determined in the laboratory does not invariably predict clinical success or failure.

Molecular Basis of Resistance

Broadly speaking, there are two main types of resistance—intrinsic and acquired. Intrinsic resistance is the inherent resistance to an antibiotic possessed by a species because of its fundamental structure or metabolism. For example, anaerobic bacteria are uniformly resistant to the aminoglycosides because they do not possess mechanisms for oxidative phosphorylation, which are required for these agents to exert an antibacterial effect.

Resistance can be acquired either by mutation of existing genetic material or by acquisition of new genetic material via transferable genetic elements, such as plasmids or transposons. Transfer of genetic material between species has great clinical relevance because it enables the spread of resistance not only within a particular bacterial species but between quite different species. This is particularly true in ITU settings, in which patients are frequently colonized by resistant bacteria, and extensive antibiotic use then leads to favorable conditions for the spread of resistance.

Mechanisms of Resistance

There are at least five main mechanisms by which bacteria exhibit resistance to antibiotics, which are shown in Table 10.4. Bacteria frequently exhibit either "cross resistance," that is, one resistance mechanism which confers resistance to several (usually related) drugs, and/or "co-resistance," in which resistance may occur to several (often unrelated) agents because of the presence of multiple resistance mechanisms in the same strain.

Why is Resistance Such a Problem in the ITU?

The risk factors for antibiotic resistance in the ITU setting are multifactorial and complex. Studies show that many patients are colonized with resistant organisms on admission to ITU.[5] Widespread antibiotic use, particularly broad-spectrum empirical regimens, gives a survival advantage to resistant strains. High patient turnover and limited space in an ITU facilitates the spread of resistant organisms into the environment and from patient to patient. Measures that help to reduce the development of antibiotic resistance in critical care are shown in Table 10.5.

TABLE 10.4. Main mechanisms of antibiotic resistance

Mechanism	Action	Example	Organism and antibiotic
Production of inactivating enzyme	Inactivation of antibiotic	ESBL	Resistance to cephalosporins in *E. coli*
Bacterial cell envelope modification	Decreased permeability of the cell to an antibiotics	OpRD porin deficiency	Resistance to carbapenems in *Pseudomonas*
Transmembrane efflux system	Antibiotics actively expelled from the cell	Tet inner membrane proteins	Tetracycline resistance in many species
Antibiotic target modification	Reduced binding of the antibiotic to its target site	Low affinity penicillin-binding protein (PBP)	β-lactam resistance in *S. aureus* (MRSA)
Acquisition of novel metabolic pathway	Antibiotic target bypassed	Reduced binding to modified cell wall precursor targets	Glycopeptide resistance in enterococci

TABLE 10.5. Minimizing the development of resistance in critical care

Action	Effect
Improved infection control measures, e.g., hand hygiene	Prevents the spread of resistant organisms between patients
Awareness of local resistant organisms	The antibiotics prescribed are more likely to be clinically effective and reduce the selection of known resistant organisms
Use narrow-spectrum antibiotics and single agents when possible	Reduces the selection of multiresistant organisms, particularly those possessing multiple resistance mechanisms
Ensure adequate doses of antibiotic are prescribed	Avoids the exposure of bacteria to subinhibitory concentrations of antibiotics, which may promote the selection of resistant strains
Avoid prolonged courses of empiric antibiotic treatment	Reduces the exposure of organisms to antibiotics unnecessarily

Current Therapeutic Problems in ITU

Methicillin-Resistant *Staphylococcus aureus*

MRSA is now a common organism, and deep-seated MRSA infections have been associated with increased mortality compared with sensitive *S. aureus* in some settings. Most strains are multiresistant, the common UK strain (known as EMRSA-15) showing resistance to erythromycin, clindamycin, and ciprofloxacin. In serious infections, therefore, therapeutic options are often limited to the glycopeptides and the newer agent, linezolid.

Coagulase-Negative Staphylococci

These skin organisms cause nosocomial infections associated with prosthetic and other indwelling devices, notably central venous cannulae. In hospitalized patients who have received previous antibiotic therapy, they are often resistant to β-lactams, gentamicin, and the quinolones. Treatment options are as for MRSA.

Enterococci (Fecal Streptococci)

These organisms are naturally sensitive to amoxicillin, but intrinsically are resistant to many antibiotics, such as the cephalosporins, quinolones, aminoglycosides, and clindamycin. They may acquire resistance to the main therapeutic options of amoxicillin, vancomycin, and teicoplanin. These GRE can only be treated with combinations of drugs or with newer agents, such as linezolid and quinupristin-dalfopristin.

Escherichia coli and Other Enteric Organisms

Forty to 60% of *E. coli* now exhibit acquired resistance to amoxicillin conferred by a plasmid-encoded β-lactamase.[1] Increasingly, *E. coli* are also carrying other resistance genes, including those encoding extended-spectrum β-lactamases (ESBLs). This confers resistance to amoxicillin and the cephalosporins and is frequently associated with co-resistance to gentamicin (15–75%) and fluoroquinolones (14–66%).[4] This often leaves limited therapeutic options, such as a carbapenem.

Other enteric gram-negative bacilli, such as *Klebsiella*, *Enterobacter*, and *Serratia*, are common nosocomial pathogens and cause outbreaks of infection, but, crucially, all can transfer plasmids encoding multiple resistance. Like *E. coli*, many strains produce ESBLs and may have other co-resistances.

Pseudomonas aeruginosa

This organism is a leading cause of nosocomial infections in the ITU and has a correspondingly high attributable mortality. It is intrinsically resistant to many antibiotics, but ceftazidime, piperacillin-tazobactam, ciprofloxacin, aminoglycosides, and meropenem are all possible treatment options. However, in more than 10% of patients, acquired resistance to these antibiotics can emerge rapidly during treatment.

Acinetobacter Species

These are hardy environmental organisms that can colonize hospitalized patients and cause opportunistic infections. They are intrinsically sensitive to a number of antibiotics, but many hospital stains are becoming multiresistant. Outbreaks of pan-resistant strains with complex resistance mechanisms have occurred in the United Kingdom.

Fungal Infections

Fungal infections in critical care patients are usually caused by *Candida* species and are an important cause of morbidity and mortality. Several factors predispose this patient group to fungemia, including previous antibiotic treatment, presence of invasive medical devices, varying degrees of immune suppression (for example, the use of corticosteroids) and comorbidities. In the persistently febrile patient receiving broad-spectrum antibiotics, it is important to have a high clinical suspicion of potential fungal infection. If *Candida albicans* is suspected, perhaps after the previous isolation of this organism from clinical specimens, fluconazole may be prescribed, but for the empirical treatment of a septic patient, a broad-spectrum antifungal, such as amphotericin or caspofungin, should be used.

Summary

The use of antibiotics in critical care is not a simple topic. Not only must the correct antibiotic be chosen for the individual patient, often taking into account polymicrobial infections with organisms exhibiting multiple antibiotic resistances, but there is an urgent need to rationalize prescribing as far as possible. This is essential if we are to retain the usefulness of the antimicrobial agents we currently have, and avoid the potential future problem of infection with bacteria that are essentially untreatable.

References

1. British Thoracic Society guidelines for the management of community acquired pneumonia in adults—2004 update.
2. Kollef MH. Optimizing antibiotic therapy in the intensive care unit setting. Critical Care 2001;5: 189–195.
3. Draft for consultation March 2005 Guidelines for the prophylaxis and treatment of MRSA infections in the United Kingdom.
4. Finch RG, Greenwood D, Ragnar Norrby S, Whitley RJ. Antibiotic and Chemotherapy. Churchill Livingstone 2003.
5. Harris AD, Nemoy L, Johnson JA, et al. Co-carriage rates of vancomycin-resistant Enterococcus and extended-spectrum beta-lactamase-producing bacteria among a cohort of intensive care unit patients. Infection Control Hosp Epidemiol. 2004;25:105–108.

Suggested Reading

British Society for Antimicrobial Chemotherapy (BSAC) website. Teaching aid on Treatment of Hospital infections. http://www.bsac.org.uk/pyxis.

British Thoracic Society guidelines for the management of community acquired pneumonia in adults— 2004 update. http://www.brit-thoracic.org.uk/bts_guidelines_pneumonia_html.

Draft for consultation March 2005: Guidelines for the prophylaxis and treatment of MRSA infections in the United Kingdom. http://www.bsac.org.uk.

Guidelines for the antibiotic treatment of endocarditis in adults: report of the Working Party of the British Society for Antimicrobial Chemotherapy. J of Antimicrobial Chemotherapy 2004;54:971–998.

Scottish Intercollegiate Guidelines Network (SIGN) for surgical prophylaxis. http://www.sign.ac.uk/guidelines/fulltext/45/index.html.

11
Infection Control in the Intensive Care Unit

David Tate and Steven J. Pedler

General Principles

Many patients admitted to critical care units carry healthcare-associated multiresistant organisms, such as methicillin-resistant *Staphylococcus aureus* (MRSA), or may have a communicable infectious disease, such as pulmonary tuberculosis (TB). Special measures are, therefore, frequently required to prevent the spread of these pathogens to other patients on the unit. In general, these are referred to as infection control procedures.

Risk to Patients

Intensive therapy unit (ITU) patients often differ greatly from patients on a general medical or surgical ward. For example, they may be intubated, have medical devices (such as intravascular cannulae or urinary catheters in situ), require frequent interventions with invasive procedures, and will often be receiving broad-spectrum antibiotics. All of these factors make the prevention of spread of potential pathogens among such patients a vital aspect of patient care in such units.

Risk to Staff

It is important to remember that infection control procedures are also designed to protect staff from infection. Healthcare staff members are usually healthy and, thus, less likely to contract infection from their patients than are patients from infected staff. However, acquisition of infection from

patients is by no means uncommon. Such infections include scabies, herpes simplex, chickenpox, TB, viral gastroenteritis, viral hepatitis, and HIV. In addition, staff may become colonized with hospital organisms without ill effects, leading to the possibility of patients becoming infected from a colonized member of staff. MRSA is perhaps the best-known example of this problem. Recommended routine infection control procedures to be followed in the care of any patient are listed in Table 11.1.

Means of Transmission of Infection

The common means of transmission of infection are by:

- The airborne route
- Fecal-to-oral transmission
- Direct or indirect contact

Typical examples of diseases transmitted by the airborne route include respiratory viruses and TB. The agents of gastrointestinal infections are excreted in feces and then must be ingested to cause infection (the fecal-to-oral route), and this includes some healthcare-associated infections, such as infections caused by *Clostridium difficile*. Opportunistic, often multiresistant pathogens, such as MRSA, may be transferred by direct contact between patients, or, more commonly, may be transmitted indirectly via the hands of nursing, medical, or other staff or via contaminated medical equipment.

TABLE 11.1. Summary of routine infection control procedures

Topic	Main points
Clinical waste	• Dispose of appropriately; do not dispose of clinical waste in domestic waste bags! • Handle sharps safely and correctly; see below • Launder linen and incinerate contaminated disposable equipment
Sharps safety	• Dispose of sharps in the appropriate container—sharps bins are provided for this purpose • Do not attempt to resheath needles, etc., which frequently leads to needlestick injuries • For their own protection, healthcare workers must be vaccinated against hepatitis B, and a range of other immunizations may be required: see your local employer's policy • Report any needle-stick injuries at once and seek appropriate advice; active and/or passive immunization, or the use of antiviral agents, may be required, and this is time critical
Decontamination and disinfection	• A clean environment is essential for prevention of healthcare-associated infection; ITU staff should not tolerate a dirty working environment • Ensure blood and body fluid spills are correctly managed; the hospital will have a policy for the correct procedure • Clean and decontaminate equipment after use; ideally personal equipment such as stethoscopes should be regularly cleaned and decontaminated (e.g., with alcohol wipes)

Hand hygiene	• Ensuring good hand hygiene is the single most important means of preventing healthcare-associated infection
	• The use of alcohol hand gel or rub is an alternative to hand washing when there is no visible soiling of the hands
	• Hand washing is *required* when there is visible soiling of the hands
	• Hand hygiene procedures must be performed:
	  ◦ on entering or leaving the unit
	  ◦ before and after contact with a patient

PPE	• This includes plastic aprons, gloves, impermeable gowns, visors or goggles, high-efficiency particle masks, etc. • PPE is worn to provide protection for the healthcare worker against infectious agents, or sometimes to prevent contamination of a very vulnerable patient (e.g., one who is very neutropenic) with organisms from staff on the unit • A risk assessment should be performed before any medical procedure and appropriate steps taken; for example, if there is a risk of aerosol formation, then the wearing of eye protection may be recommended • Cuts and abrasions on the skin of the staff member should be covered with occlusive dressings
Staff illness	• Staff with potentially infectious conditions, such as diarrhea, upper respiratory tract infections, skin sepsis, etc., should not work until given clearance to do so from their manager • If in doubt, staff should contact the occupational health department, which will advise on the staff member's fitness to work

Patient Isolation

A key factor in preventing transmission is the isolation of infected or colonized patients (sometimes incorrectly referred to as the *quarantine* of patients). Isolation facilities on an ITU may serve two very distinct patient groups:

• Source isolation: For patients who are sources of microorganisms that may spread from them and infect or colonize other patients and/or members of staff
• Protective isolation: For patients who are rendered highly susceptible to infection by disease or therapy, including immunocompromised and burn patients

The lack of isolation facilities on the ITU or the necessary staffing levels to ensure safe management of the patient is often a key hindrance to the isolation of potentially infective patients. When this occurs, a risk assessment must be performed to determine who (if anyone) should be placed in isolation. A common example would be whether a patient with MRSA or another with *C. difficile* should take priority for isolation, leaving the other patient to be nursed on the open unit. In these circumstances, the hospital infection control team (ICT) can provide help and advice.

Universal Precautions

Since the 1990s, the UK Department of Health has recommended using the same level of protection against infection with bloodborne viruses with all patients rather than attempting to identify individuals who are potential carriers of these viruses.

As a general principle, the following universal precautions should be adopted with all patients:

1. Gloves should be worn when handling blood and body fluids, e.g., amniotic fluid, pericardial fluid, peritoneal fluid, pleural fluid, synovial fluid, cerebrospinal fluid (CSF), semen, vaginal secretions, and any fluid visibly contaminated with blood.

2. Gloves should be changed after contact with every patient and hands washed immediately with hot water and soap after removing gloves.

3. If hands or other skin surfaces become contaminated with blood or body fluids, they should be washed immediately with hot water and soap and thoroughly dried.

4. Needles must not be not resheathed, purposefully bent or broken, routinely removed from disposable syringes, or manipulated by hand. Once used, sharps must be disposed of in puncture-resistant containers for appropriate disposal.

5. Masks and/or visors should be worn to prevent exposure of mucous membranes, e.g., mouth, eyes, and nose, during high-risk procedures; i.e., those that have potential to generate droplets of blood or body fluids.

6. Gowns and/or aprons must be worn during procedures that may cause splashes or sprays of blood or body fluids that soil healthcare workers clothing or uniforms.

The wearing of outdoor clothing on the ITU is not recommended. Ideally, we advocate the wearing of theater-style attire—preferably in a different color to operating theater clothing ("theater blues"). A clean set should be worn at the beginning of every shift and changed every time it becomes stained with blood or body fluids.

Specific Problems

Multiresistant Organisms

MRSA, multiresistant gram-negative bacilli (e.g., *Acinetobacter* species, *Serratia marcescens*, extended-spectrum β-lactamase [ESBL]-producing *E. coli*, and *Pseudomonas aeruginosa*), and glycopeptide-resistant enterococci (GRE) are associated with various infections on the ITU.

They are more frequently encountered in patients who have received broad-spectrum antibiotics and, once established in an ITU setting, are extremely difficult to eliminate. It should be noted, however, that some of these organisms, notably MRSA and ESBL producers, may be seen in patients admitted from the community without a prolonged hospital stay before ITU admission. Careful control of antibiotic prescribing may limit their appearance, but the strict adoption of infection control principles and procedures is crucial to prevent their spread from patient to patient.

Many critical care units screen all or selected patients for MRSA on admission by means of nose, throat, and perineal swabs. If the patient is identified as a carrier, they should be nursed in a side room if one is available. It is possible to assess the risk of MRSA carriage on admission. Patients who have had multiple previous hospital admissions, those who have received or are receiving broad-spectrum antibiotics (especially quinolones such as ciprofloxacin), those admitted from nursing homes, and those admitted from hospitals with high rates of MRSA are all at increased risk of colonization with this organism.

Surveillance of positive microbiological results from the ITU may help to identify the emergence of a particular multiresistant gram-negative bacillus. Close liaison with the microbiologist and the ICT is crucial to prevent an outbreak occurring. Often these organisms are present in the environment of the ITU and may require very thorough cleaning to remove them.

The injudicious use of antibiotics may encourage the emergence of multiresistant organisms. The development and strict implementation of an appropriate antibiotic policy will help to reduce this. However, once a patient is colonized with this type of organism, only the careful adoption of infection control procedures will prevent the spread of the organism to other patients.

A number of studies indicate that MRSA can be controlled in critical care units by the implementation of rigorous control of infection policies and protocols. For example, a "before and after" study of an ITU in a French teaching hospital with a high admission and transmission rate of MRSA reported a decrease in MRSA unit acquisition from 7.0% to 2.8% after the introduction of such a program.[1]

TABLE 11.2. Infections to be placed in respiratory isolation

Infection	Indications for ending isolation
Bronchiolitis in infants (caused by respiratory syncytial virus [RSV])	Clinical recovery. Cohort nursing of infants with RSV infection is acceptable if single-room isolation is not possible
Croup (acute laryngotracheobronchitis; most commonly caused by parainfluenza viruses)	Isolation may be appropriate in some circumstances. Discuss with the ICT
Invasive *Haemophilus influenzae* infection (meningitis, epiglottitis, etc.)	24 hours of appropriate antibiotic therapy
Influenza	Clinical recovery
Measles	5 days after the onset of rash
Mumps	9 days after the appearance of parotid swelling
Rubella	5 days after the onset of rash
Whooping cough	5 days of appropriate antibiotic therapy

TB and Other Respiratory Pathogens

Pulmonary TB represents a risk of infection to patients and staff when the sputum is smear-positive for acid-fast bacilli. The crucial first step in making a diagnosis of TB is to think of the condition, because it is no longer a common infection in the United Kingdom. With the emergence of HIV, patients may be admitted to the ITU with incidental pulmonary TB in addition to whatever condition necessitated an ITU admission. Thus, the admitting clinician must have a high index of suspicion.

Despite the recent abandonment of the childhood Bacille Calmette-Guèrin (BCG) immunization program in the United Kingdom, the use of the BCG vaccine remains an important means of preventing TB in UK healthcare staff. For this reason, every effort should be made to ensure that staff who are to work in an ITU where they may encounter patients with pulmonary TB have been immunized with BCG.

Patients suspected of having pulmonary TB should be nursed in a negative-pressure side room with a closed circuit where this is available. Patients with smear-positive disease can usually leave the side room after 2 weeks of treatment. However, the emergence of multidrug-resistant (MDR) TB presents a difficult problem, and this should be considered in the following circumstances:

- Previous incomplete or noncompliant treatment
- Contact with a patient with known MDR-TB
- Disease acquired in a country with a high incidence of MDR-TB
- Disease not responding to treatment

Advice on the infection control issues these patients present should be sought from a specialized center and isolation may be required in a dedicated isolation unit.

Other respiratory pathogens (examples are given in Table 11.2) may require respiratory isolation, the key points of which are outlined in Table 11.3.

TABLE 11.3. Key features of respiratory isolation

Aprons	Where these are normally worn for nursing procedures, they will still be required
Masks (this section applies **only** to pulmonary TB)	For the patient: not normally required; patients should be instructed to cough into tissues and to cover their mouth and nose when coughing and sneezing. All smear-positive patients should wear a mask if being transported through patient or public areas of the hospital. A high-efficiency particulate air (HEPA) mask should be worn until the patient is known not to have MDR-TB, after which, a routine surgical mask is satisfactory (which should be changed at hourly intervals).
	For staff: masks are only required when there is unavoidable exposure to respiratory secretions, e.g., during cough-inducing procedures, bronchoscopy, or prolonged care of a high-dependency patient. A mask is also appropriate when nursing a patient who is unable to cover their mouth and nose when coughing or sneezing. The ordinary type of surgical theater mask is **not** adequate for this purpose and a disposable HEPA mask must be used
Gloves	Not required unless handling blood or body fluids
Equipment	It is not necessary to have a full diagnostic kit in the room dedicated to the patient
Infective secretions	Sputum, sinus secretions, used paper handkerchiefs, and sputum cartons must be treated as infected waste and disposed of in accordance with the clinical waste disposal policy
Crockery and cutlery	No special precautions required. The patient does not require a set of dedicated utensils

Meningitis

Meningococcal meningitis can pose infection control risks to healthcare workers, especially those involved in direct airway management. *Neisseria meningitidis* is transmitted from person to person through nasopharyngeal secretions or large particle respiratory droplets that are unlikely to remain airborne beyond a distance of 1 meter.

Routine prophylaxis for healthcare workers involved in the general care of the patient is not recommended. Chemoprophylaxis is recommended only for those healthcare workers whose mouth or nose is directly exposed to infectious respiratory droplets or secretions within a distance of 1 meter from a probable or confirmed case of meningococcal disease. Very few cases of patient-to-healthcare worker transmission of meningococcal disease have been reported and the risk is, therefore, very low. The appropriate use of protective equipment when dealing with these patients, e.g., wearing masks and using closed suction, will help to reduce the necessity for chemoprophylaxis.

Pandemic Influenza

The Department of Health has produced guidelines[1] in the event of a pandemic of influenza. Pandemics occur because of the emergence of a strain of influenza that, through antigenic shift, is significantly different from previous strains. Of interest at present is the possible emergence of A/H5N1 ("Avian flu") from South East Asia. However, transmission of this strain from person to person, which is essential for a pandemic to occur, is very rare at present. Therefore, at the time of writing, a pandemic would be more likely to occur because of a more typical "human" strain of the influenza virus.

In the event of a pandemic, the Department of Health in England recommends the following measures:

- Good hand hygiene is critical to reduce the transmission of the infectious agent.
- Coughing or sneezing patients should wear surgical masks in transit to the ITU and when transferred to other units (e.g., radiology).

- Personal protective equipment (PPE) should be worn to prevent staff becoming contaminated with body fluids. Staff involved in nonaerosol-producing activities but within 3 feet of the patient should wear gloves, a plastic apron, a surgical mask, and possibly eye protection if there is a risk of splashing with respiratory secretions or other body fluids. High-efficiency respirators (such as those reaching standard EN149: FFP3) are only necessary if conducting aerosol-generating procedures (e.g., intubation, nasopharyngeal aspiration, tracheostomy care, chest physiotherapy, bronchoscopy, and nebulizer care) when a fluid-repellent gown and eye protection are also essential.
- Patients would ideally be nursed in a side room on the ITU. If none are available, it is recommended that the unit be divided into two areas; one area would nurse patients infected by influenza, preferably with their own dedicated staff.
- Disposable respiratory equipment should be used. Closed systems are preferable with the protection of a filter. Noninvasive positive-pressure ventilation should be avoided, along with water humidification.
- Only essential staff should be present when conducting aerosol-generating procedures and these should be wearing the appropriate PPE.

Severe Acute Respiratory Syndrome

The cause of severe acute respiratory syndrome (SARS) has been identified as the SARS coronavirus (SARS-CoV). Close contact with an infected person is thought to pose the highest risk of the infective agent spreading from one person to another. SARS seems to be less infectious than influenza, and the incubation period is thought to be between 2 and 7 days (maximum, 10 d).

After the worldwide concern regarding the spread of SARS, especially to the healthcare workers managing infected patients, the Health Protection Agency in England has made recommendations for the control of this infection in hospitals.[2] A summary of some key recommendations is as follows:

- The patient should be nursed either in an isolation room with negative pressure relative to the surrounding area, or a single room with its own bathroom and toilet facilities
- Protective clothing should be worn by all staff entering the room including:
  - An appropriate protective mask (such as the EN149: FFP3 respirator mask); such respirators should be worn by all personnel performing clinical care or in the room during aerosol-generating procedures
  - A long-sleeved fluid-repellent disposable gown
  - Gloves with tight-fitting cuffs for contact with the patient or their environment
  - Disposable eye protection comprising tight-fitting goggles or face shield (glasses provide inadequate protection against droplets, sprays, and splashes) during direct patient contact
- Only essential staff should enter the isolation room, and a record of all staff caring for the patient should be kept
- Patients should wear a surgical face mask, if able to do so, when in close contact with uninfected persons. The mask should be changed after 8 hours, or sooner if it becomes saturated or breathing is difficult
- If possible, aerosol-producing procedures should be avoided, but, if they are essential, they should be performed in a negative-pressure single room

Bowel and/or Enteric Organisms

Bowel or enteric organisms may pose several problems to the ITU in terms of infection control. The patient may be directly admitted from the community with severe dehydration and complications secondary to the bowel infection or may develop the infection while on the ITU. The most commonly encountered problem is likely to be *Clostridium difficile*. However, other bacterial infections, such as *Salmonella* or *Campylobacter* may occur in ITU patients, and viral gastroenteritis with agents such as norovirus or rotavirus are common, but usually self-limiting. Viral infections can, however, spread rapidly between patients and staff.

TABLE 11.4. Gastrointestinal infections

Infection	Indications for ending isolation
Campylobacter colitis	Cessation of diarrhea for 48 hours
Cholera	Cessation of diarrhea for 48 hours
Clostridium difficile-associated diarrhea or pseudomembranous colitis	Cessation of diarrhea for 48 hours
Diarrhea of unknown cause	Until communicable disease is excluded as the cause of diarrhea, or on cessation of diarrhea for 48 hours
Dysentery (*Shigella* infection or amoebic dysentery)	Cessation of diarrhea for 48 hours
E. coli gastrointestinal infection (all types)	Cessation of diarrhea for 48 hours
Gastroenteritis (viral)	Cessation of diarrhea for 48 hours
Hepatitis A	Until 1 week after onset of jaundice
Salmonella enteritis	Cessation of diarrhea for 48 hours
Typhoid or paratyphoid fever	Negative stool cultures (and urine cultures, if applicable)
Viral gastroenteritis (e.g., norovirus, rotavirus)	Cessation of diarrhea and vomiting for 48 hours

The management of these infections from an infection control viewpoint is similar. In general, however, any patient with unexplained diarrhea should be cared for in a side room. Standard isolation precautions should be performed as outlined in Table 11.1, with effective hand hygiene a vital component. Table 11.4 gives a summary of the main enteric pathogens likely to be encountered on an ITU.

Intravascular Line Infections

The use of intravascular cannulae is probably the single biggest risk factor for healthcare-associated septicemia in hospital practice today, and no less so on critical care units. Aseptic insertion and manipulation of intravascular devices is crucial, the general principles of infection control; hand washing, sterile environment, and wearing of PPE should be followed to protect both the patient and the healthcare worker. Aseptic technique does not automatically necessitate sterile gloves; a "no-touch" technique may be equally as effective with nonsterile disposable gloves for peripheral cannulae. Where the consequences of infection are increased, especially with central venous catheters, precautions to prevent infection needs to be more rigorous. Studies have shown that maximal sterile-barrier precautions; mask, sterile gown, sterile gloves, cap, and large sterile drapes are more effective than standard measures.[3,4]

The risks of acquiring an infection from an intravascular device is directly proportional to the length of time that the line/device has been in place. Removal of unneeded lines is, therefore, of great importance. Peripheral lines carry less infection risk than central lines, and substitution of peripheral for central lines should occur as soon as possible. Some studies suggest that antibiotic-coated lines reduce infection risk, but no studies have ever demonstrated efficacy using clinically relevant end points to our knowledge. Routine, prophylactic line changes have not been shown to reduce the risk of infection and are associated with increased complications. Guidelines for the prevention of catheter-related infection have been produced by the North American Center for Disease Control.[5]

Cannulae-related infection is very common in the critically ill. In the pyrexial patient who has been on the ITU for a number of days, line-related sepsis is very likely. In this situation, all lines should be removed, cultured, and replaced as needed.

Infected Staff

Healthcare workers who may be suffering from an infectious disease should refrain from work until no longer infectious. Although this sometimes poses problems in terms of staffing, this is preferable to a unit closing completely because of an outbreak. Common examples of such problems include staff with diarrhea, who should only return to work after 48 hours symptom free, upper respiratory tract infections (including the common cold, which may spread rapidly among patients and staff), and skin conditions such as eczema or psoriasis where there is heavy shedding of skin scales. Influenza vaccine should be offered to all staff within the unit. Advice on these and other issues should be discussed with both ICT and Occupational Health department.

The management of staff colonized with MRSA is controversial. At present, there is no national guidance on who to screen for MRSA and what to do for those who are found to be colonized. Our policy has been to investigate for staff carriage in a critical care unit when there have been two or more cases of MRSA infection that have a temporal link. We also have an active process of screening new staff members for carriage of MRSA.

Conclusions

In many cases, infection control is a matter of commonsense. Its aim is not to obstruct the care of patients by placing barriers in front of healthcare workers but to ensure that both staff members and patients are protected from the spread of potential pathogens. One of the simplest and most effective methods of ensuring this is the use of hand washing or the use of alcohol-based hand rubs. The importance of good hand hygiene cannot be overemphasized.

Finally, liaison with the ICT when the clinician is in doubt regarding suitable infection control procedures is *always* preferable to an outbreak on an ITU!

References

1. Lucet JC, Paoletti X, Lolom I, et al. Successful long-term program for controlling methicillin-resistant Staphylococcus aureus in intensive care units. *Intensive Care Med* 2005;31:1051–1057.
2. http://www.dh.gov.uk/PublicationsAndStatistics/Publications/PublicationsPolicyAndGuidance/PublicationsPolicyAndGuidanceArticle/fs/en?CONTENT_ID=4121752&chk=lfHfV7 (last accessed November 30, 2005).
3. http://www.hpa.org.uk/infections/topics_az/SARS/hosp_infect_cont.htm#summary (last accessed November 30, 2005).
4. Raad II, Hohn DC, Gilbreath BJ, et al. Prevention of central venous catheter-related infections by using maximal sterile barrier precautions during insertion. *Infect Control Hosp Epidemiol* 1994;15:231–238.
5. Mermel LA, McCormick RD, Springman SR, Maki DG. The pathogenesis and epidemiology of catheter-related infection with pulmonary artery Swan-Ganz catheters: a prospective study utilizing molecular subtyping. *Am J Med* 1991;91(Suppl 3B):S197–S205.
6. O'Grady NP, Alexander M, Dellinger EP, et al. Guidelines for the prevention of intravascular

catheter-related infections. Centers for Disease Control and Prevention. MMWR Recomm Rep 2002;51:1–29.

Suggested Reading

Ayliffe GAJ, Fraise AP, Geddes AM, Mitchell K. Control of Hospital Infection—A practical handbook.

Chapter 16 Special Wards and departments. Arnold (London) 2000.

Department of Health: Pandemic Influenza. A comprehensive website. http://www.dh.gov.uk/PolicyAndGuidance/EmergencyPlanning/PandemicFlu/fs/en (last accessed October 25, 2005).

Wilson J. Infection Control in Clinical Practice. Balliere Tindall (London) 2002.

12
Randomized Controlled Trials in Sepsis

Helen J. Curtis and Anna Harmar

During the past two decades, there has been a great increase in our understanding of the pathophysiology of sepsis and septic shock. This has revealed numerous potential therapeutic targets and major advances in cell and molecular biology have enabled basic scientists to synthesize a large array of new compounds and agents with a potential role in the treatment of sepsis. Using the approach outlined in Table 12.1, this chapter discusses the research basis of major clinical trials in sepsis, the evidence base for sepsis management, and highlights major issues that require further background work and clinical trials to resolve.

Management of the Proinflammatory Response

The cardinal manifestations of sepsis are caused by infection, however, many of the consequences are secondary to dysregulation of the inflammatory cascade. Numerous trials have aimed to modify this response (Wheeler 1999, Hotchkiss 2003, Annane 2005).

Anti-Tumor Necrosis Factor

Background

Tumor necrosis factor (TNF) is a ubiquitous proinflammatory cytokine and a key mediator in sepsis. In several clinical studies of sepsis, circulating TNF levels were elevated and related to increased mortality. Various strategies have been used to block TNF at the tissue level, including

TNF antibodies, and initial Phase II trials reported reductions in mortality (Abraham 2001).

Intensive Care Unit Evidence

In three randomized controlled trials (RCTs) with more than 2400 participants, there was no significant survival benefit at 28 days (Reinhart 2001, Abraham 2001, Fisher 1996).

Remaining Issues

There are a number of possible explanations for the failure of anti-TNF (and other anti-inflammatories) trials. Sepsis results in the production of numerous mediators and the relative importance of individual factors remains uncertain. Timing of intervention is important, and anti-TNF therapy may have been given too late to be effective in trials. Finally, the inflammatory response is clearly beneficial and blockage may be detrimental. This is highlighted by the increased rates of infection reported with anti-TNF therapy in rheumatoid arthritis and the increased mortality reported in a few trials of anti-inflammatories in sepsis.

Anti-Endotoxin Antibody

Background

Endotoxin administration produces signs consistent with septic shock. Monoclonal antibody technology has allowed the development of monoclonal antibodies that bind to the lipid A domain of endotoxin, and experimental studies suggested

TABLE 12.1. The pathophysiological approach to sepsis management

Management of the proinflammatory response
1. TNF antagonism
2. Endotoxin antibody
3. High-flow CVVH
Management of the metabolic response
1. Insulin
2. Corticosteroids
Management of the procoagulant response
1. Activated protein C
2. Antithrombin III
3. Tissue factor pathway inhibitor
Standard resuscitation
1. Early goal-directed therapy (EGDT)
2. Fluids
3. Choice of vasopressor

that these antibodies were protective (Cross 1994).

Intensive Care Unit Evidence

A double-blinded RCT with an anti-endotoxin antibody (Centoxin) reported a protective effect in a subset of patients with gram-negative sepsis. Survival at 28 days improved from 51 to 70% with treatment ($P = 0.014$) and improved survival to hospital discharge (63% versus 48%; $P = 0.038$) (Ziegler 1991). These encouraging results led to the granting of a European product license for this treatment.

Remaining Issues

Further analysis of the trial data indicated a number of potential problems with its design and analysis. For example, the study drug was administered to all septic patients, but subsequent analysis only involved the patients who were found to have gram-negative sepsis. The United States Food and Drug Administration (FDA) recommended further trials to prove that the benefit in one subgroup was not at the expense of another. A more robust trial showed no difference between treatment with anti-endotoxin antibody and placebo in all-cause mortality. There was even a suggestion of increased mortality in septic patients without gram-negative bacteremia receiving the monoclonal antibody (41% compared with 37% for placebo) (McCloskey 1994). This subsequently

led to the removal of the product license. The Centacor saga underlies the role of robust regulatory bodies before innovative treatments are introduced. It also highlights the difficulties and risks of conducting large commercial trials in human sepsis.

High-Volume Continuous Venovenous Hemofiltration

Background

Continuous venovenous hemofiltration (CVVH) is a potential adjuvant therapy in septic shock, in the absence of renal failure, via blood purification by removing circulating water-soluble inflammatory mediators and down-regulating the inflammatory response. Animal models (Grootendorst 1992) showed that high volume hemofiltration (6 L/h) improved hemodynamic variables. This high rate of hemofiltration would translate to a massive 12 L/h in humans.

Intensive Care Unit Evidence

Subset analysis of 40 septic patients from a RCT comparing different doses of renal replacement therapy in acute renal failure showed some survival benefit with high-volume CVVH. CVVH doses of 20, 35, and 45 mL/kg/h were investigated. At 15 days after cessation of renal replacement therapy, survivals were 25%, 18%, and 47%, respectively. The authors thought that this showed a direct correlation between treatment dose and survival (Ronco 2000, Reiter 2002). However, there are strong, theoretical arguments that challenge the concept of high-volume CVVH. CVVH is very inefficient at removing low molecular weight inflammatory mediators and the rate-limiting step in determining circulatory levels is usually the production and release of cytokines rather than the rate of removal.

Remaining Issues

Although the evidence is poor, 29% of intensive care units (ICUs) in the United Kingdom use CVVH as adjuvant therapy for sepsis routinely or often, although only one-third of these ICUs use doses higher than 4 L/min (Wright 2004).

Management of the Metabolic Response

Insulin: The Need for Tight Glycemic Control

The cellular and metabolic responses to severe sepsis include increased energy expenditure and negative nitrogen balance via changes in substrate use. However, few trials have tried to modulate the effect of the inflammatory cascade on cellular mitochondrial and metabolic activity.

Background

Hyperglycemia and insulin resistance is common in the critically ill and may contribute to complications, including infection and polyneuropathy (McCowen 2001). The Diabetes and Insulin-Glucose Infusion in Acute Myocardial Infarction (DIGAMI) trial showed that glucose control improved long-term outcome after myocardial infarction in diabetic patients (Malmberg 1999). Recent studies have tested a similar hypothesis in the critically ill.

ICU Evidence

In a landmark, partially blinded RCT, Van de Berghe et al. investigated tight glycemic control in 1548 surgical ICU patients (Van de Berge 2001). They showed that maintaining serum glucose levels between 4.4 and 6.1 mmol/L reduced ICU and hospital mortality rates compared with those with serum glucose levels between 10 and 11.1 mmol/L. Patients spending more than 5 days in the ICU showed the greatest improvement, with a reduction in mortality from 20.2 to 10.6% ($P =$ 0.005). Moreover, tight glycemic control showed significant improvement in numerous clinical morbidity outcomes.

Remaining Issues

In this trial, two-thirds of the patients were post-cardiac surgery, with mean Acute Physiology and Chronic Health Evaluation (APACHE) II score of 9 and a mortality rate of only 8% in the control group. These results may, therefore, not be applicable to the severely septic ICU population. Evidence could support the introduction of tight glycemic control to the wider ICU population because subgroup analysis showed decreased mortality from 4.2 to 1.0% in patients with a proven septic focus. Further RCTs are being conducted and these should clarify the situation in the near future.

Corticosteroids: Relative Adrenal Insufficiency

Background

Meta-analyses regarding the anti-inflammatory effect of high-dose corticosteroids in patients with septic shock showed no effect on mortality or morbidity, possibly secondary to nosocomial infections (Cronin 1995). There has been renewed interest in corticosteroid treatment after evidence of "relative adrenal suppression" in the critically ill. It has been proposed that sepsis is a state of relative adrenal insufficiency and low-dose corticosteroids could be used as adrenal replacement therapy.

ICU Evidence

A multicenter double-blinded placebo-controlled RCT compared 5 days of low-dose hydrocortisone and fludrocortisone with placebo in 300 patients with refractory septic shock. In patients with an inadequate response to short corticotrophin test (defined by a cortisol rise of <248 nmol/L, so-called nonresponders), treatment gave a 10% absolute improvement in 28-day survival ($P =$ 0.023) and reduced the duration of vasopressor therapy (Annane 2002). However, on an intention-to-treat basis for all patients (responders and nonresponders), there was no significant benefit from corticosteroids. An ongoing multicenter study with 800 patients (Corticosteroid Therapy of Septic Shock [CORTICUS]) is investigating the use of corticosteroids in nonrefractory septic shock (Annane 2005).

Remaining Issues

The diagnosis of adrenal insufficiency in critically ill patients remains a complex topic (see Chapter 3). The adrenocorticotropic hormone (ACTH) stimulation test in critically ill hypoalbuminemic patients may not be a reliable marker, because cortisol-binding globulin is reduced and bound cortisol is measured in a standard cortisol assay. The free and biologically active cortisol may be normal (Hamrahian 2004). This may be one explanation for the wide variations in results in different clinical trials. In addition, the definition of relative adrenal inefficiency remains controversial and the optimal cortisol response in severe sepsis is unknown.

Management of the Procoagulant Response

Sepsis is a procoagulant state secondary to:

1. Direct activation of coagulation by infecting organisms
2. Diffuse endothelial injury
3. Consumption of clotting inhibitors
4. Inhibition of fibrinolysis

Part of the pathophysiology of severe sepsis is thought to be secondary to unrestrictive and inappropriate coagulation in the microcirculation, and this contributes to end organ damage (see Chapter 4). The inflammation and procoagulation pathways are closely linked and synergistic. The inflammatory cytokines, TNFα, interleukin (IL)-1β and IL-6 all activate coagulation and inhibit fibrinolysis. The procoagulant, thrombin, stimulates inflammatory cascades (Hinds 2001). Coagulation is regulated by natural anticoagulants, including Protein C, antithrombin, and tissue factor inhibitors. Artificial or recombinant versions of these agents have been synthesized and used in major RCTs in sepsis.

Activated Protein C

Background

Activated protein C (APC) has a multitude of modulatory effects on the inflammatory and clotting cascade.

Anticoagulant Effects

- APC inactivates the clotting factors Va and VIIa, which prevents thrombin generation
- APC inactivates plasminogen activator inhibitor-1—a fibrinolysis inhibitor

Anti-inflammatory Effects

- APC reduces IL6 and other cytokine synthesis
- APC inhibits platelet activation, neutrophil recruitment, and mast cell degranulation

Reduced levels of APC have been noted in sepsis, and are associated with worse mortality (Yan 2001). Studies have shown that APC infusions lead to reduced levels of d-dimers and IL-6, indicators of effects of coagulation and inflammation, respectively (Bernard 2001).

ICU Evidence

The Recombinant Human Activated Protein C [Xigris] Worldwide Evaluation in Severe Sepsis (PROWESS) multicenter double-blinded placebo-controlled RCT (1690 participants) compared an infusion of APC with placebo in septic ICU patients. The trial was halted early after a second interim analysis suggested the treatment effect was large. The results showed a 6.1% absolute reduction in 28-day mortality with APC ($P = 0.006$) (Bernard 2001). There was a small increased bleeding risk; 3.5% in the treatment group versus 2.0% with placebo. Intracranial bleeds were seen in 0.2% of treated patients. A recent open-label trial of APC, which included UK patients, reported a similar mortality to the treatment limb of the PROWESS study. However, the study did not contain a control group (Wheeler 2003).

Remaining Issues

The use of APC as a supplementary therapy in patients with severe sepsis and two-organ failure has been endorsed by various clinical bodies (including the US FDA, European drug-regulating bodies, and the UK National Institute for Clinical Excellence [NICE] agency). However, a number of issues regarding the use and efficiency of APC remain:

1. Follow-up data of the original PROWESS patients beyond 28 days do not report significant improvements in outcomes at 3, 6, 12, or 24 months (Angus 2004).
2. Administration of APC to less severely ill patients (i.e., single organ failure) has not proven to be beneficial and may even be detrimental (Abraham 2005).

3. In non-RCT settings, the intracranial hemorrhage rate has been reported as high as 1.5%
and serious bleeding in the Abraham study was
2.5% (Wheeler 2003).

APC may confer both benefits and risks to
patients in a manner similar to warfarin therapy
in stroke prevention. This could explain the findings that only the more severely ill patients gain
benefit from treatment. However, the possibility
that the benefit of APC in the PROWESS study
arose by chance cannot now be completely dis-
missed. It has been suggested that further RCTs
of APC should be performed to resolve these
issues.

Antithrombin III

Background

Antithrombin (AT)-III acts as an anticoagulant
via its binding and, thus, inactivation of thrombin, factors IXa, Xa, and XIa. This reaction is
increased 1000-fold in the presence of heparin.
Meta-analysis of experimental studies suggested
that supraphysiological doses of ATIII were protective against mortality and multiorgan failure
(MOF) in sepsis (Eisele 1998).

ICU Evidence

In the KyberSept double-blinded placebo-
controlled RCT of ATIII 2339 septic patients were
randomized to treatment or placebo. No survival
difference was seen between the two groups at 28
days (Warren 2001).

Remaining Issues

Two important issues arose in the study:

1. Heparin administration plus study drug
increased 28-day mortality from 36.6 to 39.4%
($P = 0.02$).
2. The 680 patients who did not receive heparin
during the treatment days had an improvement in
survival at 28 and 90 days with ATIII compared
with placebo (37.8% versus 43.6%, $P = 0.08$; and
44.9% versus 52.3%, $P = 0.03$; mortality rates at 28
and 90 d, respectively).

These findings have rekindled an interest in
the effects of heparin in sepsis and a number of

studies are currently underway to explore these
issues.

Tissue Factor Pathway Inhibitor

Introduction

Tissue Factor is the major initiator of blood coagulation. Tissue Factor Pathway Inhibitor (TFPI) is
synthesized by endothelial cells and blocks coagulation in its earliest phase via factor Xa inhibition
and prevention of factor VII activation. Early
Phase II trials of exogenous TFPI suggested an
improved outcome in sepsis (Abraham 2000).

ICU Evidence

The Optimized Phase III Tifacogin in Multicenter
International Sepsis trial (OPTIMIST) study
double-blinded placebo-controlled RCT investigated TFPI versus placebo in 1754 severely septic
patients with mild coagulopathy (international normalization ratio [INR] ≥ 1.2). No effect was seen on
mortality rates at 28 days (34.2% versus 33.9%;
treatment versus placebo) (Abraham 2003).

Remaining Issues

The trial raised several interesting issues:

1. At the interim analysis of the first 722
patients, the survival difference in favor of TFPI
was almost sufficient to activate the predefined
stopping rules (29.1% versus 38.9% mortality
rates in treatment and placebo groups). The trial
continued with subsequent treatment-allocated
patients showing a significant decline in survival.
Outcome in the PROWESS study also differed significantly when the first and second halves of the
trial were compared. In the PROWESS study,
treatment benefit was only observed during the
second half of the study.
2. A parallel study group of 201 patients with
low INR (<1.2) showed a lower mortality in TFPI-
treated patients (12% versus 22.9%). Although
not of statistical significance, this may reveal a
potential target subgroup.
3. Subgroup analysis of the treatment group
found no difference in survival between those
receiving or not receiving concomitant heparin.
However, in the placebo group, those who received
heparin had a lower mortality rate (29.8% versus

42.7%; $P = 0.05$). It is clear that the role of heparin in sepsis requires further investigation.

Standard Resuscitation

Early Goal-Directed Therapy

Background

The concept of "early goal-directed therapy" (EGDT) involves prompt identification of patients at risk of progression to septic shock. It enables target-driven treatment in the "golden hours" and aims to optimize cardiac preload, afterload, and contractility, thus, preventing the consequences of oxygen debt.

ICU Evidence

River's landmark RCT with 263 patients investigated EGDT in the first 6 hours of hospital admission (River 2001). Treatment-group patients received in sequential fashion:

1. Fluid resuscitation
2. Vasopressor or dilator agents
3. Red blood cell transfusion
4. Inotropic treatments

The goals of treatment were (Figure 12.1): 1) a central venous pressure (CVP) of 8 to 12 cmH$_2$O; 2) a mean arterial pressure (MAP) of 65 to 90 mmHg; 3) urinary output greater than 0.5 mL/kg/hr; 4) a mixed venous oxygen saturation (ScvO$_2$) greater than 70%; and 5) a hematocrit of greater than 30%. Standard therapy did not include the final two goals.

Of treated patients, 99.2% achieved these goals within 6 hours compared with 86% of control patients who received standard care. There was a significant reduction in hospital mortality in the treatment group (30.5% versus 46.5; $P = 0.009$) and less organ dysfunction (APACHE II scores in EGDT group, 13.0 ± 6.3 versus 15.9 ± 6.4 in controls; $P < 0.001$).

Remaining Issues

To translate this study into clinical practice will require effective collaboration between front-line clinicians and critical care departments regarding early recognition and aggressive management of septic patients.

Fluids: Crystalloid Versus Colloid

A recent and comprehensive Cochrane review on fluid resuscitation in the critically ill, using all-cause mortality as an end point, investigated numerous colloids compared with crystalloids. These included albumin or plasma protein fraction (19 trials, 7576 participants), starch (10 trials, 374 participants), gelatin (7 trials, 346 participants), and dextran (9 trials, 834 participants). It concludes that there is currently no evidence to support one type of fluid regime (Roberts 2004).

Albumin has traditionally been used as a resuscitation fluid. A Cochrane review (Cochrane 1998) suggested that administration of albumin of any concentration resulted in a 6% increase in overall mortality. However, a recent double-blinded RCT, which compared 4% albumin with normal saline resuscitation, reported similar mortality and morbidity outcomes for both groups (The Safe Study Investigators 2004). In view of these findings, the use of albumin as a resuscitation fluid has been abandoned by most intensive therapy unit (ITU) practitioners.

Choice of Vasopressor

1. Standard vasopressors: the role of vasopressors in patients with sepsis is undisputed in the face of circulatory failure despite fluid resuscitation. The choice of vasopressor, however, remains unclear. A Cochrane review could only identify seven RCTs (total patients, 172) and concluded that no agent was superior, although none of the trials used patient-orientated outcome measures (Muller 2005).

2. New agents: exogenous vasopressin is increasingly used in the treatment of septic shock. In contrast to other vasopressors, in low doses, it has vasodilatory effects on coronary, cerebral, and pulmonary circulations (Holmes 2001).

ICU Evidence

Normal plasma vasopressin levels are less than 4 pg/mL in overnight fasted, hydrated humans.

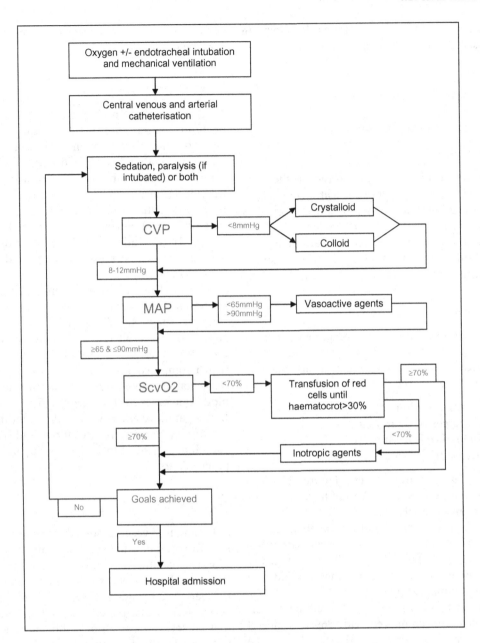

FIGURE 12.1. Protocol for EGDT. (Rivers, 2001.)

Hypotension and hypovolemia are potent stimuli that exponentially increase vasopressin levels. However, in advanced vasodilatory septic shock, plasma levels of vasopressin are low compared with cardiogenic shock, and vasopressin administration in septic shock increases arterial blood pressure (BP) (Landry 1997). Other trials are currently being undertaken. Vasopressin administration may be considered a physiological replacement rather than pharmacological therapy (Malay 1999). A small RCT in vasopressor dependant septic shock showed that vasopressin at 0.04 U/min increased systolic BP from 98 to 125 mmHg ($P < 0.05$) and enabled withdrawal of all other catecholamines.

The trials to date are small and do not investigate the effects of vasopressin on clinical outcomes, such as organ dysfunction and survival.

Conclusions

In the last 20 years, Critical Care practitioners and the pharmaceutical industry have conducted several large and well-conducted RCTs. Disappointedly, many of these trials have not shown any benefit from the investigational agent. However, recent studies of APC, insulin, and early goal-directed resuscitation suggest some room for optimism. There are no shortages of potential therapeutic targets in sepsis (rather the opposite!). However, it is probably naive to think that any single agent will have a large impact on survival and it is likely that progress will occur by achieving small, but cumulative, increments in survival.

References

Abraham E. Tissue factor inhibition and clinical trial results of tissue factor pathway inhibitor in sepsis. *Crit Care Med* 2000;28(suppl 9):S31–S33.

Abraham E, et al. Lenercept in severe sepsis and early septic shock. *Crit Care Med* 2001;29:503–510.

Abraham E, Laterre PF, Garg R, et al. Drotrecogin Alfa (Activated) for Adults with Severe Sepsis and a Low Risk of Death. *N Engl J Med* 2005;353:1332–1341.

Abraham E, Reinhart K, Opal S, Demeyer I, et al. Efficacy and safety of Tifacogin (recombinant Tissue factor pathway inhibitor) in severe sepsis. *JAMA* 2003;290:238–247.

Angus DC, Latterre PF, Helterbrand J, et al. The effect of doltrecogin alfa (activated) on long-term survival after severe sepsis. *Crit Care Med* 2004;32: 2199–2206.

Annane D, Bellissant E, Cavaillon JM. Septic shock. *Lancet* 2005;365:63–78.

Annane D, Sebille V, Charpentier C, Bollaert PE, et al. Effect of treatment with low dose hydrocortisone and fludrocortisone on mortality in patients with septic shock. *JAMA* 2002;288:862–871.

Bernard GR, Vincent JL, Laterre PF, LaRosa SP, Dhainaut JF, Lopez-Rodriguez A, Steingrub JS, Garber GE, Helterbrand JD, Ely EW, Fisher CJ Jr. Efficacy and safety of recombinant human activated protein C for severe sepsis. *N Engl J Med* 2001; 344(10):699–709.

Cochrane Injures Group Albumin Reviewers. Human albumin administration in critically ill patients: systemic review of randomized controlled trials. *BMJ* 1998;317:235–240.

Cronin L, Cook DJ, Carlet J, et al. Corticosteroids treatment for sepsis: a critical appraisal and metaanalysis of the literature. *Crit Care Med* 1995;23:1430–1439.

Cross AS. Antiendotoxin antibodies. *Ann Int Med* 1994;121:58–60.

Dellinger RP, Carlet JM, Masur H, et al. Surviving sepsis campaign guidelines for management of severe sepsis and septic shock. *Intensive Care Med* 2004;30:536–555.

Eisele B, Lamy M, Thijs LG, et al. Anti-thrombin III in patients with severe sepsis. *Intensive Care Med* 1998;24:663–672.

Fisher CJ, Agosti JM, Opal SM, Lowry SF, et al. Treatment of septic shock with tumour necrosis factor receptor:Fc fusion protein. *N Eng J Med* 1996;344: 1697–1702.

Gattinoni L, Brazzi L, Pelosi P, et al. A trial of goal-orientated hemodynamic therapy in critically ill patients. *N Engl J Med* 1995;333:1025–1032.

Grootendorst AF, van Bommel EFH, van der Hoven B. High volume hemaofiltration improves hemodynamics of endotoxin-induced shock in the pig. *J Crit Care* 1992;7:67–75.

Hamrahian AH, Oseni TS, Arafah BM. Measurements of serum free cortisol in critically ill patients. *N Engl J Med* 2004;350:1629–1638.

Hinds CJ. Treatment of sepsis with activated protein C. *BMJ* 2001;323:881–882.

Holmes CL, Patel BM, Russel JA, Walley KR. Physiology of vasopressin relevant to management of septic shock. *Chest* 2001;120:989–1002.

Hotchkiss RS, Karl IE. The pathophysiology and treatment of sepsis. *N Engl J Med* 2003;348: 138–150.

Landry DW, Levin HR, Gallant EM, et al. Vasopressin deficiency contributes to vasodilation of septic shock. *Circulation* 1997;95:1122–1125.

Malay MB, Ashton RC, Landry DW, et al. Low-dose vasopressin in the treatment of vasodilatory septic shock. *J Trauma* 1999;47:699–703.

Malmberg K, Norhammar A, Wedel H, Ryden L. Glycometabolic state at admission: important risk marker of mortality in conventionally treated patients with diabetes mellitus and acute myocardial infarction: longterm results from the Diabetes and Insulin-Glucose infusion in Acute Myocardial Infarction study (DIGAMI). *Circulation* 1999;99:2626–2632.

McCloskey RV, Straube RC, Sanders C, Smith S, et al. Treatment of septic shock with human monoclonal antibody HA-1A. *Ann Int Med* 1994;121:1–5.

Mullner M, Uranek B, Havel C, et al. Vasopressors for shock. The Cochrane database of systematic reviews 2004, Issue 3. Art. No: CD003709.pub2. DOI: 1002/14651858.CD003709.pub2.

NICE Guideline—www.nice.org.TA084Sepsis (severe)—drotrecogin—Guidance.

Reinhart K, et al. for the AFELIMOMAB Sepsis Study Group. Randomized-placebo-controlled trial of the anti-tumour necrosis factor antibody fragment Afelimomab in hyperinflammatory response during severe sepsis: RAMES Study. *Crit Care Med* 2001; 29:765–769.

Reiter K, Bellomo R, Ronco C, Kellum J. Pro/con clinical debate: Is high-volume haemofiltration beneficial in the treatment of septic shock? *Crit Care* 2002; 6(1);18–21.

Rivers E, Nguyen B, Havstad S, et al. Early goal-directed therapy in treatment of severe sepsis and septic shock. *N Engl J Med* 2001;345:1368–1377.

Roberts I, Alderson P, Bunn F, et al. Colloids versus crystalloids for fluid resuscitation in critically ill patients. The Cochrane Database of Systemic Reviews 2004, Issue 4. Art No.:CD000567.pub2.DOI: 10.1002/14651858.CD000567.pub2.

Ronco C, Bellomo R, Homel P, Brendolan A, Dan M, Piccinni P, La Greca G. Effects of different doses in continuous veno-venous haemofiltration on outcomes of acute renal failure: a prospective randomised trial. *Lancet* 2000;356:26–30.

The Safe Study Investigators. A comparison of albumin and saline for fluid resuscitation in the intensive care. *N Engl J Med* 2004;350:2247–2256.

Van de Berghe G, Wouters P, Weekers F, et al. Intensive insulin therapy in critically ill patients. *N Engl J Med* 2001;345:1359–1367.

Warren BL, Eid A, Singer P, Pillay SS, et al. High dose Antithrombin III in severe sepsis. *JAMA* 2001; 286:1869–1878.

Wheeler AP, Bernard GR. Treating patients with severe sepsis. *N Eng J Med* 1999;340:207–214.

Wheeler AP, Doig C, Wright T, et al. Baseline characteristics and survival of adult severe sepsis patients treated with drotrecogin alfa (activated) in a global, single-arm, open-label trial (ENHANCE). *Chest* 2003; 124:91S, abstract.

Wilson J, Woods I, Fawcett J, et al. Reducing the risk of major elective surgery: randomised controlled trial of preoperative optimisation of oxygen delivery. *BMJ* 318:1099–1103.

Wright SE, Bodenham A, Short AIK, Turney JH. The provision and practice of renal replacement therapy on adult intensive care units in the United Kingdom. *Anaesthesia* 2003;58(11):1063–1069.

Yan SB, Helterband JD, Hartman DL, et al. Low levels of protein C are associated with poor outcome in severe sepsis. *Chest* 2001;120:915–922.

Zeigler EJ, Fisher CJ, Sprung CL, Straube RC, et al. Treatment of gram-negative bacteremia and septic shock with HA-1A human monoclonal antibody against endotoxin. *New Eng J Med* 1991;324: 429–436.

13
Guidelines, Protocols, and the Surviving Sepsis Guidelines: A Critical Appraisal

Ian Nesbitt

In 2004, the Society for Critical Care Medicine and the European Society of Intensive Care Medicine published the "Surviving Sepsis Campaign Guidelines" (SSCG) (Table 13.1). These were endorsed by a number of international organizations and, since publication, there has been a sustained and ongoing campaign to encourage adoption of the SSCG by critical care practitioners worldwide.

The aim of this article is to outline some issues in guideline development with reference to these guidelines. A detailed critique of each component of the SSCG is beyond the scope of this article (see further reading), but it is worth using the SSCG for examples of some of the difficulties inherent in guideline development. The focus of this article will be on United Kingdom practice.

What is the Purpose of Guidelines?

As defined by the American Institute of Medicine, clinical guidelines are "systematically developed statements to assist practitioner and patient decisions about appropriate healthcare for specific clinical circumstances." Most clinical practice guidelines (CPGs) are developed to improve healthcare by bridging the gap between research and daily practice. Protocols are more circumscribed, typically being a set of directions on actions to be taken, with little allowance for clinical judgement.

Ideally, guidelines should make clinical practice easier by assimilating a wealth of relevant information and producing it in a readily available, easy to understand and use format. Unfortunately, this is more the exception than the rule, and the majority of an ever-expanding number of CPGs fail even simple appraisal tools (such as the Appraisal of Guidelines Research and Evaluation [AGREE] instrument; Table 13.2).

Why do Guidelines Fail?

Although most guidelines are drawn up with current best practice in mind, critics argue that they reduce all clinical standards to the average and impair freedom to manage individual patients. There are numerous reasons why the process of implementing guidelines fails (Table 13.3), discussed in more detail below.

Evidence is Lacking

Gathering robust evidence in critical care is difficult. To study a sufficiently large, homogenous group of patients requires multicenter studies, prolonged study periods, or both. This introduces bias into the results because the patients may be managed differently between centers, or those at the end of the study period may be managed differently from those at the start. By the time a particular trial has been published, clinical practice may have changed significantly, possibly even making the initial question irrelevant. For example, the control group in the Acute Respiratory Distress Syndrome (ARDS) Network (ARDSnet) trial is regarded by some as having been enrolled into a "worse than standard level of

TABLE 13.1. A brief overview of the surviving sepsis campaign guidelines

Area	Recommendation	Grade of evidence
Initial Resuscitation	Use Early Goal Directed Resuscitation	B
Infection—Diagnosis, source control and antibiotic therapy	Early identification, removal of any focus of infection, start antibiotics within 1 hour of diagnosis	D/E
Vasopressor Use	Norepinephrine/dopamine as first choice. Vasopressin as second line	D/E
Inotropic use	Dobutamine if low cardiac output after fluid resuscitation	E
Low dose Steroids	2–300 mg/day for vasopressor dependant patients	C
Blood Products	Transfusion trigger of 7 g/dl unless acute coronary syndrome/acute anaemia.	B
	Platelets if platelet count $<5 \times 10^9$/l (higher if risk factors)	E
Sedation/analgesia/neuromuscular blockade	Avoid neuromuscular blockade if possible.	E
	Use sedation protocols.	B
Mechanical Ventilation	Use low tidal volume ventilation (6 ml/kg lean body weight)	B
Blood Glucose Control	Aim for glucose <8.3 mmol/l	D
Activated Protein C	Use for patients with high risk of death (APACHE II score >24, sepsis induced multi-organ failure)	B
Stress Ulcer prophylaxis	Use H2 antagonists	A
Thromboembolism prophylaxis	Use heparin +/− mechanical devices	A
Bicarbonate therapy	Not indicated to improve haemodynamics or hypoperfusion induced acidosis pH > 7.15	C
Renal Replacement	CVVH and IHD are equivalent	B

Grading of Evidence.
A—supported by at least two level I investigations.
B—supported by one level I investigation.
C—supported by level II investigations only.
D—supported by at least one level II investigation.
E—supported by level IV or V evidence only.

care" arm. In some areas of medicine, adequate randomized trials are extremely difficult to design and perform (e.g., pregnancy and pediatrics). The overall result is that many practices in critical care medicine lack good quality evidence to support (or oppose) them.

TABLE 13.2. The AGREE instrument (appraisal of guidelines for research & evaluation)

AGREE instrument domain	Overview
1. Scope & purpose	What is the overall aim of the guideline?
2. Stakeholder involvement	Does the guideline represent the views of the people who will use it?
3. Rigour of development	Is the process of developing the guideline robust?
4. Clarity & presentation	Is the guideline easy to read & use?
5. Applicability	What are the implications (organisation, cost & behaviour) of adopting the guideline?
6. Editorial independence	Any conflicts of interest in producing the guideline?

Available Evidence Is Misinterpreted or Misleading

Many guideline developers and busy clinicians are unable to assimilate all relevant publications (including unpublished research) to adequately interpret the quality of all relevant evidence, so rely on information filtered by experts, editors, and clinicians with particular interests. This is an imperfect approach, adding to any bias inherent in research actually available.

TABLE 13.3. Common reasons why guidelines fail

1. Adequate evidence is lacking
   • No studies in appropriate area
   — Available evidence is misinterpreted or misleading
   — Guidelines are not applicable to patients outside studies
2. Expert beliefs or misconceptions influence published guidelines
3. Guidelines lack consistency
4. Hasty recommendation/adoption of expensive interventions diverts resources from other areas of care
5. Official guidelines are accepted at face value
6. Guidelines are too inflexible or complex to manage individual patients

Some trials with a large impact on clinical practice are seriously flawed (because of poor study design, bias, statistical errors, etc.), but their shortcomings are revealed late, or are inadequately publicized, by which time, clinical practice has altered, sometimes irrevocably. Examples include the reduction in pulmonary artery catheter (PAC) use after the Study to Understand Prognoses and Preferenced for Outcomes and Risks of Treatment (SUPPORT) study, and of albumin after the Cochrane collaboration meta-analysis in 1998.

Even when the data are certain, recommendations for or against interventions will involve subjective value judgments when the benefits are weighed against the harms. Those making the judgements may have different value systems from those affected by the decisions.

Guidelines Are Not Applicable to Real-Life Patients

Many studies enrol a highly select group of patients. Real-life patients may be sufficiently dissimilar to these patients to invalidate the findings. Recombinant human activated protein C (rhAPC) is currently recommended for septic patients with a high risk of death, although licensing details vary between countries. The original trial, Recombinant Human Activated Protein C [Xigris] Worldwide Evaluation in Severe Sepsis (PROWESS), has been followed by several other studies (Extended Evaluation of Recombinant Human Activated Protein C [ENHANCE], Administration of Drotrecogin Alfa (Activated) in Early Stage Severe Sepsis [ADDRESS], and Researching Severe Sepsis and Organ Dysfunction in Children: A Global Perspective [RESOLVE]). These studies have raised sufficient concern regarding rhAPC that the editorial accompanying the ENHANCE trial states that: "At least one prospective randomized controlled trial should be conducted to confirm that rhAPC is effective in those patients with sepsis for whom it is currently recommended." At the time of writing, Eli Lilly has just announced that this will take place during the next 2 years. Although the current SSCG for the use of rhAPC are in accord with the outcomes of studies to date, it is likely that further refinement of which patient groups benefit from the drug will

alter future published recommendations. The parallels with Centoxin (a human monoclonal antibody against endotoxin, withdrawn from use in the early 1990s) are disturbing.

Numerous studies exist to show that protocol-driven weaning from mechanical ventilation is faster than and as safe as that directed by physicians. The SSCG advocate daily trials of spontaneous breathing. Much of the evidence to support this comes from healthcare settings in which the model of care is dissimilar to that in the United Kingdom (open versus closed intensive care unit [ICU] organization). The overall evidence in support of weaning protocols is much less robust than its component parts would suggest, and improving ventilator technologies may circumvent some of the historical requirement for guidelines.

Tight blood sugar control for critically ill patients was defined as 4.4 to 6.1 mmol/L in Van den Berghe's landmark study (in mainly postcardiac surgical patients). Subsequent results from more heterogenous patient groups have reinforced the clinical benefits of normoglycemia, although they have also raised concerns regarding frequent hypoglycemic events. The SSCG recommendation of a target blood glucose level of less than 8.3 mmol/L are a compromise between these ideal levels, the risks of hypoglycemia, and what is feasible in a typical ICU environment. Again, ongoing large-scale studies may better inform future SSCG recommendations.

Expert Beliefs or Misconceptions Influence Recommendations

The assumed benefits from the use of PACs were unchallenged for many years, in part because of Consensus statements from experts despite limited evidence of benefit. Likewise, low-dose steroid replacement for septic shock is a currently recommended practice that carries significant caveats. The SSCG recommend steroid replacement for patients with shock (Level C evidence). Unfortunately, the definition of adequate adrenal response during critical illness is still unclear, the dose of synacthen used for testing is debatable (as is the requirement for any adrenal testing), and the values used to assess adequacy of response are also not fully agreed. Opinion within the critical care community ranges from "corticosteroids should be considered for all patients with septic shock" to "There is no evidence to base a treatment strategy on cortisol levels."

Interestingly, Selective Decontamination of the Digestive Tract (SDD) is not mentioned in the SSCG, yet has evidence of at least Grade C (some argue Level A) to support it. The SSCG authors state that they did not include SSD because it is an "infection prevention" treatment, rather than a "diagnosis and management" of sepsis intervention. This is despite including other "prevention" strategies, such as stress ulcer prophylaxis (SUP) and thromboembolism prophylaxis in the SSCG. SDD is an area of critical care medicine subject to much argument, and perhaps the topic was simply too controversial to achieve a consensus recommendation.

Guidelines Lack Consistency

Guidelines should have internal and external consistency. Many guidelines contradict themselves and each other, leading to ambiguity in interpretation. Examples in the SSCG include variable advice regarding the use of dobutamine, and extrapolation of information (on glycemic control, SUP, and thromboembolic prophylaxis) from sepsis trials to nonseptic patients and vice versa. Another example, regarding 20 different guidelines for the management of atrial fibrillation in Northern England lack congruity to the extent that only 1% of patients would receive the same treatment under every guideline.

Hasty Recommendation/Adoption of Expensive Interventions Diverts Resources from Other Areas of Care

rhAPC is expensive (estimated cost £80 million/yr in England and Wales). Compared with some chemotherapy regimes, rhAPC is a cost-effective treatment, but, to date, world sales of rhAPC have been well below those predicted. This may reflect clinicians rationing treatment, or being reluctant to use the drug on current evidence, despite official recommendations. It is unclear what effect the use of rhAPC has on other areas of healthcare in the United Kingdom.

Official Guidelines Are Accepted at Face Value

The SSCG recommend H2 antagonists as the agents of choice for SUP (rated as Grade A evidence). The evidence regarding SUP has not kept pace with changes in clinical practice. The need for SUP seems to be declining (for many reasons). Newer agents (proton pump inhibitors [PPI]) have not yet been used in adequately sized trials to compare them with traditional H2 blockers and sucralfate. Thus, although H2 blockers may be rated as the SUP agents of choice for septic patients, based on the results of trials performed principally in the 1980s, many clinicians would disagree that this is the most appropriate evidence base from which to base current practice. This is reflected in a recent study showing marked increases in the use of PPI in critical care.

Additionally, practices that are suboptimal from the patient's perspective may be recommended to help control costs, serve societal needs, or protect special interests (e.g., those of doctors, risk managers, or politicians). The short-lived introduction of disposable instruments for adenotonsillectomy in the United Kingdom in 2001 as a response to fears regarding prion disease is an example of this approach.

Guidelines Are Too Complex or Inflexible for Daily Use

Even if a treatment is regarded widely as effective, actually translating research findings into routine practice may be extraordinarily difficult. Low tidal volume ventilation has been increasingly practiced for at least a decade, and was widely acclaimed after the publication of the ARDSnet trial in 2000. However, many centers struggle "in real life" to achieve the goals set down in this trial. Similar difficulties have been reported in adopting sedation and glycemic control guidelines. The significant time, energy, and administrative burdens involved are contributing factors to modest uptake of guidelines.

Guideline/Protocol Development

It is important that physicians who use guidelines are aware of the process and difficulties associated

TABLE 13.4. Developing guidelines and protocols

1. Select & Prioritise a Topic
2. Set up a Team
3. Involve Patients & Users
4. Agree Objectives
5. Build awareness & commitment
6. Gather Information
7. Baseline Assessment
8. Produce the Protocol
9. Pilot the Protocol
10. Implement the Protocol
11. Monitor Variation
12. Review the Protocol

with guideline production. For those involved with guideline development and leading multidisciplinary teams, this knowledge is even more important.

The National Health Service (NHS) modernization agency suggests a 12-point process (Table 13.4).

Conclusion

Many guidelines fail to achieve their purpose, but good guidelines have the potential to rapidly disseminate "best practice." It is likely that initiatives such as the Leapfrog group (in the United States) and payment by results (in the United Kingdom) will use adherence to guidelines as an indicator of quality care in the future. This means that guidelines are unavoidable for most clinicians. Uncritical adoption of published clinical guidelines is misguided, but equally so is hypercritical dissection of existing guidelines and clinical evidence. Guideline development is more complex than it seems at first sight. Clinicians reading, reviewing, or developing guidelines should be aware of the complexities involved in translating research findings into evidence-based clinical practice.

Anyone developing guidelines should adopt a structured approach, both to the process of writing each guideline, and to the overall framework in which the individual guidelines sit. Even well constructed, appropriate guidelines are infrequently followed by practicing clinicians, so one should avoid unrealistic expectations about what they will accomplish. In an area of constant change, any guideline is likely to become rapidly dated, and the evidence base on which it is built equally so.

Overall, this means that the SSCG should be used as a framework on which to build, using detailed knowledge and critical appraisal to formulate appropriate, workable local clinical guidelines, rather than as protocols or expected standards of care.

Suggested Reading

Surviving Sepsis Campaign Articles

Beale R, Singer M. Surviving Sepsis Guidelines—Bundles to Adopt? JICS. 2005;6(3):14–16.

Daley RJ, Rebuck JA, Welange LS, et al. The changing demographics of stress ulcer prophylaxis: Prevention of stress ulceration: current trends in Critical care. Crit Care Med. 2004;32(10):2008–2013.

Dellinger RP, Carlet JM, Masur H, et al. Surviving Sepsis Campaign Guidelines for management of severe sepsis and septic shock. Background, Recommendations, and Discussion from and evidence based review. Crit Care Med. 2004;32(11 Suppl).

Discussion of the evidence for steroids in sepsis: http://bmj.bmjjournals.com/cgi/eletters/329/7464/480 (accessed October 18, 2005).

Landucci D. Critique of SSCG: The surviving sepsis guidelines: Lost in translation. Crit Care Med. 2004;32(7):1598–1600.

van Saene HK, Petros AJ, Ramsay G, et al. All great truths are iconoclastic: selective decontamination of the digestive tract moves from heresy to level 1 truth. Int Care Med. 2003;29(5):677–690.

Viviani M, Silvestri L, van Saene HK, Gullo A. Surviving Sepsis Guidelines: Selective decontamination of the digestive tract still neglected. Crit Care Med. 2005; 33(2):462–463.

Young MP, Manning HL, Wilson DL, et al. Ventilation of patients with acute lung injury and acute respiratory distress syndrome: has new evidence changed clinical practice? Crit Care Med. 2004;32(6):1260–1265.

General Protocol/Guideline Articles

A guide to protocol development: Key steps to developing protocols. NHS Modernisation Agency/NICE. www.srht.nhs.uk/docs/guidance/step2step.pdf (accessed October 18th 2005).

Hammond JJ. Protocols and guidelines in critical care: development and implementation. Curr Op Crit Care 2001;7(6):464–468.

The AGREE tool: http://www.agreecollaboration.org/ (accessed October 18, 2005).

Woolf SH, Grol R, Hutchinson A, et al. Potential benefits, limitations, and harms of clinical guidelines. BMJ. 1999;318:527–530.

Other Articles

Butler R, Keenan SP, Inman I, et al. Is there a preferred technique for weaning the difficult to wean patient? A systemic review of the literature. Crit Care Med. 1999;27(11):2331–2336.

Cochrane Injuries Group Albumin Reviewers. Human albumin administration in critically ill patients: systematic review of randomised controlled trials. BMJ. 1998;317:235–240.

Connors AF, Speroff T, Dawson NV, et al. The effectiveness of right heart catheterization in the initial care of the critically ill. For the SUPPORT Investigators. JAMA. 1996;276:889–897.

Guidelines for atrial fibrillation: Evidence based guidelines? Bandolier. 102. http://www.jr2.ox.ac.uk/bandolier/band102/b102-2.html. Accessed October 19, 2005.

Sackett DL. Levels of evidence: Rules of evidence and clinical recommendations on the use of antithrombotic agents. Chest. 1989;95(2):S2–4.

14
Practical Approaches to the Patient with Severe Sepsis: Illustrative Case Histories

Victoria Robson

The purpose of this chapter is to show how therapeutic strategies and evidence from various studies can be best brought together and prioritized when treating patients with life-threatening sepsis. Much needs to be achieved in a short space of time, and prioritizing can be difficult. In this situation, diagnosis and treatment cannot be separated. Rapid action plans need to be made, often on the basis of limited information and then revised repeatedly based on further investigation and treatment response. "Do no harm" is the first rule of medicine, but in severe sepsis, "doing nothing" invariably leads to a poor patient outcome.

The management starts with a rapid assessment of the patient, targeting assessment and emergency management of the airway, breathing, and circulation so that compromised airway, inadequate breathing, and inadequate circulation are addressed immediately.

Then follows a systematic search for the source of the sepsis and degree of derangement of physiology, which includes history taking, patient examination, and study of the case notes, so that a more specific management plan can be established. The management will include further resuscitation, organ support, and treatment of the underlying cause of the problem, in addition to consideration of the most appropriate placement of the patient (e.g., critical care unit).

The initial management plan should include the following:

- High-flow facemask oxygen (or ventilatory support if necessary).

- Intravenous access, preferably wide-bore (16 or 14 gauge) cannula inserted into a peripheral vein for fluid resuscitation.
- Continuous monitoring using pulse oximetry, three-lead continuous electrocardiography (ECG) and automated noninvasive blood pressure (BP) measurement, repeating at 3-minute intervals.
- Arterial cannulation, for intra-arterial BP measurement, which is particularly useful when BP is low (noninvasive measurement may be inaccurate) and when inotropes or vasopressors are needed, and also for repeated blood sampling.
- Central venous cannulation, for delivery of inotropes or vasopressors, measurement of central venous pressure (CVP), and measurement of central venous oxygenation.
- Intravenous fluid, preferably by bolus "fluid challenge" using a targeted end point such as CVP measurement. It is likely that both crystalloids and colloids are equally effective, provided sufficient volume is given.
- Obtaining microbiological samples, including blood for culture and any other relevant sample (urine, sputum, pus, cerebrospinal fluid [CSF], as appropriate). Ideally, these should be obtained before the first dose of antibiotic is administered, unless the time taken to obtain the specimen is such that treatment would be delayed. An example of this would be if meningitis is suspected, antibiotic administration should not be delayed while a lumbar puncture for CSF is obtained; however, blood cultures should be obtained as intravenous access is

achieved, immediately before administering the antibiotic.

- A plan may be needed for urgent invasive procedures to sample infected sites or provide therapeutic strategies. Examples would include ultrasound or computed tomographic (CT)-guided aspiration of a collection of pus, laparotomy for investigation and treatment of peritonitis, or surgery to debride an area of soft tissue gangrene. The balance of risk (moving an unstable patient to an area of the hospital where resuscitation may be more difficult) and benefit (obtaining a microbiological sample or limiting the spread of infection) needs to be considered carefully in each case.
- Appropriate antibiotic; "best guess" will need to be used, based on clinical suspicion of site of infection, knowledge of likely organism(s), and knowledge of the sensitivity pattern of these likely organism(s) modified by the local situation of antibiotic resistance and prescribing policies.
- Assessment of adequacy of organ perfusion. This will include clinical assessment for signs such as confusion, oliguria, hypotension, metabolic acidosis, elevated serum lactate, and may also include invasive monitors, such as pulmonary artery catheter measurement of cardiac output and oxygen delivery.
- Cardiovascular support for inadequately perfused organs. This may include intravenous fluid, red cell transfusion, and vasoactive medication. Rapid restoration of adequate organ perfusion is vital to prevent organ failure.
- Consideration of the most appropriate location for the patient, e.g., transfer to critical care unit, and arrangement for a safe journey.

After the initial period of resuscitation, reassessment is essential, with further treatment as required. Particular attention should be paid to:

- Organ perfusion. Continued vigilance as to fluid status is needed. As the capillary bed becomes "leaky," there may be loss of intravascular fluid into the interstitium, causing hypovolemia. Organ perfusion may be further compromised by excessive vasodilatation and/or myocardial depression, each of which needs appropriate treatment. The choice of drug to improve organ perfusion depends on the

clinical circumstance; a hypotensive patient with a vasodilated peripheral circulation (low systemic vascular resistance), warm feet, and high cardiac output, is likely to benefit from noradrenaline infusion, at an infusion rate titrated to effect, and, if necessary, by the addition of vasopressin infusion. However, when cardiac output is low, then an inotrope such as dobutamine should be considered. Use of a cardiac output monitor (such as pulmonary artery catheter, esophageal Doppler probe, lithium dilution cardiac output [LiDCO], or pulse contour cardiac output [PiCCO]) may assist in these decisions—however, there is no evidence to support the use of any one monitor over the others.

- Control of blood glucose concentration. Short-acting insulin infusion should be commenced to control blood glucose to be within the range of 4.4 to 6.1 mmol/L.
- Steroids. Evidence supports the use of 200 mg/d hydrocortisone either by 24-h infusion or in four divided doses. Some would advocate performing a short synacthen test before steroid administration.
- Revision of antibiotic therapy, guided by microbiological advice, when responsible organism(s) and organism sensitivities are known, and by clinical response.
- Early nutrition, by enteral route whenever possible. This usually entails passage of an enteral feeding tube (e.g., nasogastric tube) with continuous complete enteral feeding.
- Support of any organs that fail, e.g., renal support if acute renal failure occurs.
- Use of activated protein C (drotrecogin-α) when appropriate

Illustrative Cases

Example 1

A 50-year-old man with a history of gallstones, and recent cholecystitis, who is on a waiting list for cholecystectomy, presents with a 2-day history of right upper quadrant abdominal pain, tachycardia, and pyrexia. He has been treated with cefuroxime and paracetamol. A critical care referral is made when his systolic BP is noted to be 70 mmHg, with heart rate of 145 beats per minute and temperature of 40°C.

Priorities

Assess and manage airway, breathing, and circulation:

- Airway is patent (patient is talking but confused).
- Respiratory rate is 35 breaths per minute, without focal clinical signs; saturation is 99% on high-flow oxygen.
- Tachycardia and hypotension, with warm peripheries, suggesting septic shock, requires wide-bore peripheral intravenous access (e.g., 14 gauge) and commencement of fluid resuscitation (start with 500 mL gelofusine rapid infusion).

Although arterial and central venous cannulation will be needed (arterial for accurate beat-to-beat measurement of BP, which may inaccurately measured by noninvasive methods when hypotensive, and central venous for guiding fluid replacement and administration of inotropes), these are not a priority.

If BP continues to fall despite rapid fluid infusion, then a vasopressor such as noradrenaline should be commenced, if necessary, via peripheral cannula until central intravenous access is obtained. Risk of extravasation should be balanced against clinical need.

Sepsis Screen

A search for the source of sepsis, to guide investigations and antibiotic therapy, should then occur; the clinical findings of right upper quadrant tenderness, along with the history, here suggest biliary tract sepsis. Blood cultures should have been obtained already (from the peripheral cannula as it was inserted). Investigations to obtain a sample of infected material could include ultrasound- or CT-guided drainage of the biliary tract. In this case, the patient's cardiovascular condition means it would be safer to bring ultrasound to him rather than transport him to the radiology department; this should be arranged immediately and should occur in the next hour, if possible.

Broad-spectrum antibiotics, which are appropriate for biliary tract organisms, should be administered; a suitable choice would be piperacillin-tazobactam, or the addition of gentamicin to cefuroxime. The first dose needs to be administered as soon as the blood cultures are taken, and should not wait for radiology-guided aspiration.

Reassessment and Further Management

After initial management, the patient is reassessed for signs of improvement or deterioration, particularly in respiratory and cardiovascular systems, and further resuscitation is directed by this assessment. The initial aim is to stabilize the patient sufficiently to transfer to the critical care unit.

There invasive lines can be inserted (if not already done); this is also a suitable opportunity to obtain blood for lactate measurement, central venous oxygenation, and other routine tests (white cell count, clotting, etc.). Ultrasound-guided aspiration can be performed and the sample obtained sent for urgent gram stain and culture. A fresh pus sample is placed into a sterile container (not onto a swab). Urinary catheterization should be performed.

Example 2

A 70-year-old man, who has spent 19 days in intensive therapy unit (ITU) after an anterior resection for sigmoid carcinoma, has become drowsy, hypotensive, and pyrexial. The surgery had been complicated by a perioperative myocardial infarction and prolonged weaning from ventilation. The patient is receiving assisted spontaneous ventilation of 20 cmH$_2$O inspiratory pressure with 5 cmH$_2$O positive end-expiratory pressure (PEEP) via a tracheostomy, and receiving 50% oxygen, with a respiratory rate of 28 breaths per minute.

Arterial blood gases demonstrate: PO_2, 11.1 kPa; PCO_2, 6.3 kPa; base excess, −6 mmol/L; lactate, 4.5 mmol/L; pH, 7.25. Observations include a heart rate of 110; BP, 80/50 mmHg; CVP, +8 mmHg; temperature, 38.5°C; and cool peripheries.

Initial Resuscitation

- Airway: Already secured (tracheostomy).
- Breathing: The presence of a lactic acidosis implies tissue hypoxia caused either by inadequate oxygen delivery to vital organs or inability of cells to consume oxygen. In this situation, increasing the inspired oxygen will have little impact on tissue oxygenation, although it does give an increased margin of safety if gas exchange worsens. If PCO_2 or respiratory rate rise, he may need a change of ventilation mode to include mandatory breaths.
- Circulation: One or more fluid challenge is needed, guided by change in CVP.

Sepsis Screen

The source of the sepsis should be sought, as well as consideration of nonseptic causes of shock (e.g., concealed bleeding or a cardiogenic problem).

Possible septic sources could include:

- Anastomotic breakdown (late, but healing may have been impaired by the perioperative myocardial infarction).
- Respiratory system (ventilator-associated pneumonia or aspiration pneumonia).
- Cannula-related sepsis.
- Wound infection.
- Pressure sore.
- Urinary tract infection.

Cultures are needed of blood (peripheral "stab"), blood from central line, sputum, urine, swabs from the abdominal wound and any other wounds (e.g., pressure sore), and consideration given to imaging the abdomen.

If the central venous cannula is considered as a possible source of sepsis, it should be removed, and replaced by a new one at a different location. The tip of the removed catheter is sent for microbiological culture.

A broad-spectrum antibiotic should be commenced. If the clinical source of sepsis is not apparent, then the choice will need to be suitable for all possibilities (to include skin organisms, hospital-acquired respiratory organisms, and gut organisms), as well as taking into account any recent positive cultures and the likelihood of resistance of organisms to any recently used antibiotics. The possibility of fungal infection should also be considered, particularly if multiple courses of antibiotics have already been administered. Advice from a senior microbiologist should be obtained.

If the BP and organ perfusion do not respond to fluids, then the clinical picture of cold peripheries and recent myocardial infarction would support the use of dobutamine. A cardiac output monitor would be useful to guide this choice.

Red cell transfusion should also be considered as part of the initial resuscitation. The patient is likely to have a hemoglobin concentration of 7 to 9 g/dL (having been in the ITU for 19 d). If oxygen delivery is low, particularly as further hemodilution occurs, then transfusion may be useful.

Example 3

A 66-year-old woman with mild osteoarthritis and hypertension presents with a 2-day history of dry cough, increasing shortness of breath, and general malaise. She has been noted to be tachypneic, with a respiratory rate of 40 breaths per minute, and pulse oximetry saturation of 75% on air, increasing to 84% on facemask oxygen (reservoir mask with 15 L/min oxygen flow). Auscultation demonstrates quiet breath sounds in the right lower zone with an area of bronchial breathing and dullness to percussion. Heart rate is 130 beats per minute; BP, 125/60 mmHg. The following results have been obtained:

Hb, 12.2 g/dL; wbc, 1.3 × 10⁹ cells/L; platelets, 743 × 10⁹ cells/L

Na, 142 mmol/L; K, 5.7 mmol/L; urea, 23.7 mmol/L; creatinine, 327 μmol/L

Blood glucose, 12.3 mmol/L

Arterial blood gases (on oxygen): pH, 7.01; PO_2, 6.5 kPa; PCO_2, 6.8 kPa

Base excess, −10 mmol/L; lactate, 5.4 mmol/L

Rapid Assessment

- Airway: Patent.
- Breathing: Inadequate rapid shallow breathing with hypoxemia and Type 2 respiratory failure.
- Circulation: Tachycardia, normotension (but low diastolic BP), with warm peripheries in a restless, agitated patient.

Resuscitation Priorities

Management of the inadequate breathing (identified by the low saturation, low PO_2, and raised PCO_2); this lady is tiring and approaching a respiratory arrest. Because she is already receiving maximal facemask oxygen, facemask continuous positive airway pressure (CPAP) might be tried as a "holding measure" while equipment, drugs, and wide-bore peripheral intravenous access are prepared for intubation and ventilation. Intubation may be hazardous and should, therefore, be performed by the most experienced and skilled practitioner available; preoxygenation will have little effect on PO_2, and rapid desaturation should be expected on apnea. Initially ventilation should be with 100% oxygen—VQ matching worsens after induction of anaesthesia.

A fall in BP should also be anticipated; the elevated serum lactate reflects shock (inadequate tissue perfusion) as well as tissue hypoxemia, and the apparently normal BP may be maintained by the patient's own elevated catecholamine levels. Reduction of these after anesthesia may reveal the degree of hypovolemia. A rapid fluid challenge should *precede* induction of anesthesia, and inotrope/vasoconstrictors, such as adrenaline and noradrenaline, should be immediately available. Sputum should be obtained for culture after intubation.

After securing the airway and establishing mechanical ventilation, the next priority is to optimize the circulation. Peripheral cannulation will have been established to induce anaesthesia. This should be used for repeated fluid challenges, using BP and jugular venous pressure to guide therapy, until central venous access is achieved. If hypotension persists after adequate intravenous fluid, then noradrenaline infusion should be started. The choice of drug is guided by the clinical situation; the warm vasodilated peripheries with rapid capillary refill along with the chest signs of consolidation and pyrexia suggest septic shock with excessive vasodilation, making noradrenaline a suitable first-line therapy.

At insertion of the central venous catheter line, blood cultures should be taken, along with the sputum cultures taken earlier. Serum for atypical pneumonia serology and urine for *Legionella spp.* antigen should also be obtained. The choice of antibiotic is directed by the clinical situation—community-acquired pneumonia seems to be the most likely cause of the sepsis. Factors should be sought in the history and clinical findings that would predispose to particular types of pneumonia; e.g., does the patient have an underlying respiratory condition or susceptibility to particular infections (e.g., undergoing chemotherapy), has she recently been abroad or in contact with other chest sepsis patients, are there abnormalities of liver function tests or hyponatremia to suggest *Mycoplasma pneumoniae* infection?

A combination of antibiotics, such as cefuroxime and erythromycin, would be a suitable choice for such a community-acquired pneumonia. This choice can be modified after microbiological confirmation of the responsible organism.

After intubation and circulatory resuscitation, this patient should be transferred to the intensive care unit, to commence renal replacement therapy, along with insulin infusion to control her blood glucose level to within the range 4.4 to 6.1 mmol/L, 200 mg hydrocortisone infusion per 24 hours, nasogastric tube insertion for enteral nutrition, and administration of activated protein C (unless contraindications exist).

Practical Tips and Avoiding Pitfalls

- Young patients with sepsis are much sicker than they superficially look. Always check their base excess and lactate.
- Tissue hypoxia can occur in the absence of hypoxemia. Check for lactic acidosis.
- Shock can occur in the absence of hypotension. Check for lactic acidosis.
- Do not delay intubation in severe sepsis until respiratory arrest has occurred. Intubation is usually needed.
- Commence fluid resuscitation before starting to insert invasive cannulae—the insertions will take longer than you expect.
- Do not resuscitate only to a target BP—organ perfusion may still be compromised despite adequate BP, particularly when vasoactive drugs are used.
- Do not assume that "pulmonary edema" reflects fluid overload—the same signs can also be

caused by noncardiogenic acute respiratory distress syndrome (ARDS), in a potentially hypovolemic patient.

- Take blood (and other) cultures before starting antibiotics. The presence of antibiotics in the culture sample may inhibit the bacteria and prevent their identification.
- Check that the antibiotics have been administered (not just prescribed) or administer them yourself.
- Patients who develop sepsis after abdominal surgery commonly have an abdominal source of infection. Do not diagnose a postoperative "chest infection" until an abdominal source has been excluded.
- Patients with "normal" abdominal CT scans can still have an abdominal source of sepsis.
- If it is drainable, drain it.
- Venous catheter-related sepsis is common. "If in doubt, pull it out."
- Reexamine the patient—purpura and subcutaneous gas may lie beneath the sheets.
- Talk to the family. They need information and may mention the recent trip to Africa.

Conclusions

Septic shock presents one of the greatest challenges to critical care practitioners. There is now conclusive evidence that rapid, appropriate, and targeted resuscitation can improve patient outcome. The team effort and personal skills that this requires are good markers of a high-quality intensive care unit.

References

Dellinger RP, Carlet JM, Masur H, et al. Surviving Sepsis Campaign guidelines for management of severe sepsis and septic shock. *Intensive Care Med* 2004;30:536–555.
Rivers E, Nguyen B, Havstad S, et al: Early goal-directed therapy in the treatment of severe sepsis and septic shock. *N Engl J Med* 2001;345:1368–1377.

Index

A

ABCs (airway, breathing, and circulation), of sepsis management, 93, 96
Acinetobacter, antibiotic-resistant, 68–69
Adrenal insufficiency, relative, 13, 80–81
Adrenocorticotropic hormone (ACTH), 12, 15
Adrenocorticotropic hormone stimulation test, 12–13, 80–81
Albumin, 83, 89
American Institute of Medicine, 87
Amikacin, 64, 66
Aminoglycosides, 64, 66
Amoxicillin, 63, 66
  microbial resistance to, 69
Anemia, 19, 57
Antibiotics, 63–69, 94. *See also names of specific antibiotics*
  broad-spectrum, 65, 95, 96
  microbial resistance to, 67–68
  prophylactic use in surgical patients, 45
Antibody deficiency, 47–48
Anticoagulant therapy, 22–23, 81–83
Anti-endotoxin antibody therapy, 78–79
Antifungal agents, 50–51, 69
Antithrombin III, 19, 22, 82
Anti-tumor necrosis factor therapy, 78
Appraisal of Guidelines Research and Evaluation (AGREE), 87, 88

Artemisinin, 58
Arthritis, β-hemolytic streptococcal, 42–43
Aspergillosis, invasive pulmonary, 50
Atelectasis, as postsurgical fever cause, 45
Atrial fibrillation, 90
Avian influenza, 56, 74

B

Bacille Calmette-Guèrin (BCG) immunization, 73
Bacteremia, 42, 49, 66
Beta-1 adrenergic agonists, 35–36
Bloodborne viruses, universal precautions for, 71–72
Blood cultures, for sepsis evaluation, 96, 98
Blood flow, in sepsis, 33–35. *See also* Microcirculation
Blood glucose control. *See* Glycemic control, tight
Blood pressure, intra-arterial measurement of, 93
Blood products
  cytomegalovirus-negative, 53
  as disseminated intravascular coagulation treatment, 20–22
  use in thrombocytopenic patients, 18
Bone marrow transplant recipients, cytomegalovirus reactivation in, 53
Bronchiectasis, 51

C

Calcineurin inhibitors, 51–52
Campylobacter, as gastrointestinal infection cause, 75
Candida albicans, 69
Carbapenems, 63–64, 69
Carbimazole, as granulocytopenia cause, 48
Cardiovascular resuscitation, 35–36
Cardiovascular support, 35, 94
Cardiovascular system, effect of sepsis on, 32–35
Carditis, β-hemolytic streptococcal, 42–43
Catheters
  central venous, 49, 93, 96
  pulmonary artery, 89, 94
  as sepsis cause, 45–46, 98
  as urinary tract infection cause, 45–46
  venous, as sepsis cause, 98
CD14, genetic polymorphisms in, 29
Cefotaxime, 63, 66
Ceftazidime, 63
Cefuroxime, 66, 94, 95
Cell-mediated immunity, defects in, 48–49
Cellular abnormalities, in sepsis, 37
Cellulitis, streptococcal, 42
Centacor, 79
Centoxin, 78, 89
Central venous pressure (CVP), 93
Cephalosporins, 63
Cerebrospinal fluid (CSF) analysis, 41, 93–94

Chlamydia pneumoniae, 65
Cholera, 75
Ciprofloxacin, 66, 69
Clinical waste, proper disposal of, 71
Clostridium difficile, 65, 66, 71, 72, 75
Coagulation, in sepsis, 19–23, 81
  genetic polymorphisms affecting, 28
Coagulopathy, "consumption," 20
Colloids, as resuscitation fluids, 83, 93
Colony-stimulating factors, 17
Complement deficiency, terminal, 40
Coronavirus, as severe acute respiratory syndrome (SARS) cause, 58, 74
Corticosteroids
  immunosuppressive effects of, 51–52
  as septic shock treatment, 80–81
Cortisol, 12–13, 15, 90
Cryoprecipitates, as disseminated intravascular coagulation treatment, 21
Crystalloids, as resuscitation fluids, 83, 93
Cyclooxygenase, 7
Cyclosporin A, 51–52
Cytokines
  genetic polymorphisms in, 29–30
  in hyperglycemia, 12
  in innate immune response, 5
  in multiorgan dysfunction syndrome (MODS), 32
Cytomegalovirus infections, 49, 53

D
Disseminated intravascular coagulation, 18, 19–23
Dobutamine, 90, 94, 96
Drotrecogin, 22–23, 94

E
Elderly patients, sepsis in, 2–3, 4
Eli Lilly, 89
Endocrine response, to critical illness and sepsis, 11–15
Endothelins, in multiorgan dysfunction syndrome (MODS), 32

Endothelium, vascular, 32, 36, 38
Endotoxin tolerance, 9–10
Enteral nutrition, 94
Enteric organisms, infection control of, 75
Enterococci, antibiotic-resistant, 69, 72
Epstein-Barr virus infections, 49, 52
Erysipelas, streptococcal, 42
Erythromycin, bacterial resistance to, 69
Escherichia coli infections, 49, 69
Etomidate, 13, 15
European Society of Intensive Care Medicine, "Surviving Sepsis Campaign Guidelines" of, 87–92
Exotoxins, streptococcal, 42

F
Fasciitis, necrotizing, 42, 44–45, 66
Fibrinogen, 20, 21, 23
Filtration, high-efficiency particulate air (HEPA), 50
Flucloxacillin, 63
Fluid resuscitation, 83, 93
Fluoroquinolones, 50, 64
Fungal infections, 4, 48, 49, 69

G
Gangrene, 44
Gastrointestinal infections, 71, 75
Gastrointestinal tract, selective decontamination of, 90
Genetics, of sepsis and inflammation, 26–31
  genetic polymorphisms, 26–30
  terminology of, 31
Gentamicin, 64, 66, 95
Glomerulonephritis, β-hemolytic streptococcal, 42–43
Gloves, protective, 71, 72
Glycemic control, tight, 12, 15, 80, 89
Glycopeptides, 64–65
Gram-negative bacterial infections
  as granulocytopenia cause, 48, 49
  in hospitalized patients, 1
  multidrug-resistant, 72
  treatment of, 63–64, 79

Gram-positive bacterial infections, as granulocytopenia cause, 48, 49
Granulocyte colony-stimulating factor, 51
Granulocytopenia, 47, 48, 49–51
Growth hormone, 14
Guidelines, clinical, 87–92
  failure of implementation of, 87–91
  purpose of, 87

H
Haemophilus influenzae, 51, 65, 66
Hand washing, as infection control method, 45, 50, 71
Healthcare workers
  infection prevention in, 70
  infectious diseases in, 75
Hematological changes, sepsis-related, 17–19
Hemofiltration, high-volume continuous venovenous (CVVH), 79
Hemorrhage, disseminated intravascular coagulation-related, 19, 20, 21
Heparin, 18, 22–23, 82
Herpes simplex infections, 52
Histamine, 5
Histamine$_2$ antagonists, 90
Human immunodeficiency virus (HIV), polymorphism-related inhibition of, 27
Human immunodeficiency virus (HIV) infection, 48, 51, 53–55, 70, 73
Hydrocortisone, 13, 15, 94
Hyperglycemia, 12, 15, 80
Hypogammaglobulinemia, 40, 51
Hypogonadism, 14–15
Hypoperfusion, hepatomesenteric, 35
Hyposplenism, as meningococcal sepsis risk factor, 40
Hypothalamo-pituitary-adrenal axis, 12–13
Hypothalamo-pituitary-gonadal axis, 14–15
Hypothalamo-pituitary-thyroid axis, 13–14
Hypoxemia, intracellular, 37
Hypoxia, 36